THE PORTSMOUTH DOCKYARD STORY

THE PORTSMOUTH DOCKYARD STORY

FROM 1212 TO THE PRESENT DAY

PAUL BROWN

The
History
Press

Cover images: *Front*: The launch of the battleship *Queen Elizabeth*; The aircraft carrier *Queen Elizabeth* berthed at Portsmouth; HMS *Warrior* and the dockyard seen from Spinnaker Tower; The launch of the battleship *Inflexible*, 1876; The submarine *Onyx* in No. 12 Dock (PRDHT). *Rear*: *Queen Elizabeth* berths at Portsmouth Dockyard's Princess Royal Jetty for the first time.

Frontispiece: HMS *Dreadnought* was a potent symbol of the impressive capabilities of Portsmouth Dockyard in the Edwardian era, having run her first steam trials just a year and a day after her keel was laid. (R. Silk)

First published 2018

Reprinted 2020, 2025

The History Press
97 St George's Place,
Cheltenham, Gloucestershire, GL50 3QB
www.thehistorypress.co.uk

British Library Cataloguing in Publication Data.
A catalogue record for this book is available from the British Library.

ISBN 978 0 7509 8602 1

Typesetting and origination by The History Press
Printed in Italy by Elcograf Spa

CONTENTS

◇

The past is the only dead thing that smells sweet.

Edward Thomas (1878–1917), 'Early one morning in
May I set out', 1917.

◇

From where did the might of the Navy spring? From the broad oaks in ancient forests, felled by the woodman's axe; from the teams of horses which hauled the oak to the dockyard; from the sawyer's toil as he sided the timber; from the skilful strokes of the shipwright's adze; from the exquisite plans carefully drafted by the master shipwright; from the hands of the rope maker, sail maker and rigger – the artisans who fashioned the hemp and tar and canvas of the great ship's rig. Such was the work of the dockyard in the days of wood and sail, with unchanging skills being handed down from generation to generation. The aromas of freshly worked oak, elm and pine, and of pitch, tar and hemp, hung in the air. The skeleton of a ship-of-the-line towered above the building slip as the work of framing and planking the hull progressed, and all around were stacks of timber which was being left to season, whilst planks were being soaked in steam chests so that they could be carried hot to the ship's side and bent to shape against the frames. In the mast houses, the floors were lined with newly cut sections of spruce masts, bowsprits and yards. In the smith's shop, sparks flew from red-hot metal as the iron hoops for the masts and other fittings were fashioned. Dry docks built of wood or masonry, in which the ships' bottoms could be cleaned, repaired and sheathed, provided a focal point for the teeming activity of shipwrights, caulkers, oakum boys, painters, apprentices and scavelmen. It was a man's world, save for the Colour Loft where women were sewing flags, whilst in the Sail Loft the sewing was done by men. Amidst all the toil and tar there were a few oases of calm – the gardens which graced the grand house of the commissioner and those of the dockyard officers, which also had a fish pond and an avenue of lime trees; even the porter's lodge had a garden.

The Navy was of paramount importance to the nation, for as King Charles II's Articles of War in the First Dutch War proclaimed, 'It is upon the navy under the good providence of God that the safety, honour and welfare of this realm do chiefly depend.' And the dockyards provided vital support to the Navy – building, fitting out, repairing and rebuilding the ships. From the late seventeenth century until the early twentieth century, Portsmouth Dockyard was the largest industrial organisation in Great Britain, and probably the largest in the world.

The first dockyard that we know of at Portsmouth was founded by King John in 1212, probably on a site used by his predecessor, King Richard I, to berth ships from 1194. But it was to be relatively short-lived: damaged by storms in 1228, it never fully recovered and by 1253 had been abandoned. The present dockyard site – the oldest in the country – is further up harbour, and was founded in 1495 by King Henry VII. It remained the most important dockyard throughout Henry VIII's reign, but then went into relative decline until its marked resurgence under Cromwell's Commonwealth in the mid seventeenth century. By 1698 it was again the most important dockyard in the country, and a programme of expansion and renewal was under way. In the middle of the eighteenth century, an ambitious plan for further expansion and development of the dockyard was in place, creating many of the Georgian buildings and docks still seen in the Historic Dockyard. Under the impetus of the American, French Revolutionary and Napoleonic wars, the yard continued to grow, and had a workforce of 4,300 by 1813.

Between the First Dutch War (1652–54) and victory in the Napoleonic Wars (1815), the King's Navy evolved to a position of superiority over all rival navies. Thereafter, until the entry of the United States into the Second World War, Britain's Royal Navy held indisputable command over the oceans of the world.

The nineteenth century was marked by continuous technological change which transformed the Navy from wood and sail to one of steam and iron, with equally transformational changes in the dockyards. The operations of the yard were increasingly mechanised, new skills and technologies were used to construct and repair ships, and marine engineering became an essential part of activities. Unlike their counterparts in private shipyards, the dockyard shipwrights adapted to the change from wood to iron and steel, and were complemented by new trades such as boilermakers, rivet boys, millwrights, engine smiths, pattern makers and hammer men. The expansion of the empire and Navy demanded vast new dockyard facilities, and two massive extensions to the yard were built, mostly on land newly reclaimed from the harbour. Here sprung up foundries, with blinding streams of white-hot metal flowing from ladles into moulds, a great smithery, with huge steam hammers sending off thousands of brilliant sparks in every direction, and factories to make a thousand parts for engines and equipment; powerful steam engines were installed to drive the dock pumps, machine tool workshops and metal mills. Now the skyline was dotted with the tall chimneys of boiler houses, from which drifted the smoke of the Industrial Revolution. And from the building slips a cacophony arose from hundreds of hammers striking a million rivets into the plates of the latest steel leviathan.

In the Edwardian era, Portsmouth became the leading yard for construction of the revolutionary new Dreadnoughts. By 1914, all the extension work in the yard was complete and during the First World War workforce numbers in the yard topped 25,000, the highest ever. Throughout the twentieth century, the dockyard remained relatively unchanged physically, and the demands of another world war were met. However, the Navy experienced a continuous reduction in size after the Second World War, which meant that such extensive dockyard facilities were no longer needed. In 1984, the yard was dramatically downsized and lost its status as a royal dockyard. Much of the redundant part was opened to the public as the Historic Dockyard, which is now a destination for 750,000 visitors each year, attracted by the best-preserved Georgian dockyard in the world which, uniquely, is complete with a magnificent contemporary warship, HMS *Victory*, and simultaneously allows sights of the modern Royal Navy's ships.

Some of the historic buildings of the dockyard are outside the heritage area to which the public has access, despite them being disused, redundant to the Navy's needs. Of these, a number are at risk, in a state of decay, ravaged by dry rot or water penetration. The best solution would be for these buildings to be repaired and assigned other uses, as part of an extended Historic Dockyard and the National Museum of the Royal Navy, with increased public access. Though the cost of remedial work is high, these buildings cannot be left to rot – they are part of not only Portsmouth's heritage but also the nation's. Within their walls lie hundreds of years of history, and imaginative ideas for their reuse can be found.

This book presents the social, organisational, architectural, technological and naval history of the dockyard in a period covering 800 years. From muddy creek to naval-industrial powerhouse; from building wooden walls to building Dreadnoughts; from King John's galleys to the enormous new Queen Elizabeth-class aircraft carriers, this is the story of Portsmouth Dockyard.

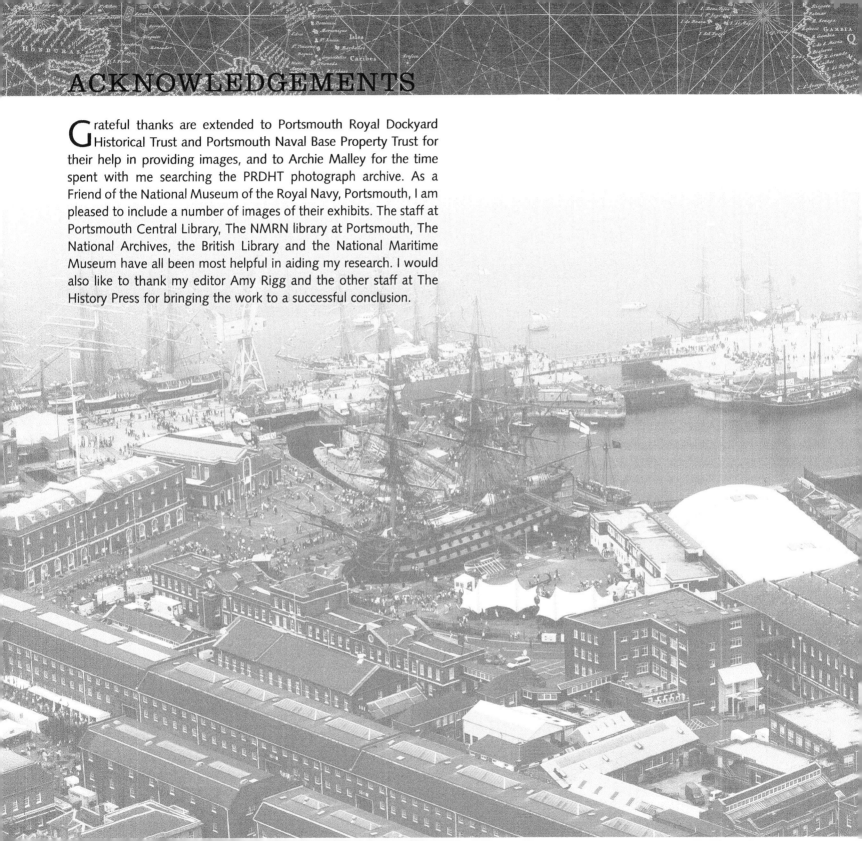

ACKNOWLEDGEMENTS

Grateful thanks are extended to Portsmouth Royal Dockyard Historical Trust and Portsmouth Naval Base Property Trust for their help in providing images, and to Archie Malley for the time spent with me searching the PRDHT photograph archive. As a Friend of the National Museum of the Royal Navy, Portsmouth, I am pleased to include a number of images of their exhibits. The staff at Portsmouth Central Library, The NMRN library at Portsmouth, The National Archives, the British Library and the National Maritime Museum have all been most helpful in aiding my research. I would also like to thank my editor Amy Rigg and the other staff at The History Press for bringing the work to a successful conclusion.

We order you, without delay, by the view of lawful men, to cause our Docks at Portsmouth to be enclosed with a Good and Strong Wall ... for the preservation of our Ships and Galleys.

King John to the Sheriff of Southampton, 20 May 1212

Portsmouth as a city or town has long been defined by its dockyard, whose history can be traced back over 800 years. The natural sheltered harbour there has for many centuries provided a haven for the Navy and a port of embarkation for successive armies. It is complemented by the anchorage at Spithead which, like the harbour entrance, is shielded by the Isle of Wight from the prevailing south-westerly winds. These advantages have led to 2,000 years of maritime activity there, and to Portsmouth's role as Britain's most important naval port during crucial periods in its history.

The Romans built a stronghold at Portchester in the third century as part of a chain of forts from Brancaster (in Norfolk) to Portchester, and this site was possibly the one named Portus Adurni. Building activities probably began in the late third century around the time when the corrupt naval commander Carausius seized independent power in Britain, and he or his successor, Allectus, may have been responsible for constructing the fort at Portchester.[1] It has the most complete Roman walls in Northern Europe. They are 20ft thick, and the front face contains bastions which accommodated ballista (Roman catapults). By AD 501, the Roman occupation had declined and Portchester may have been used by the Saxons as a defence against Viking attacks, until some control was gained by the ships of King Alfred and his successors in the late ninth and tenth centuries.

King John's Dockyard

Henry I built a castle within the Roman walls at Portchester and embarked from Portsmouth on several occasions for Normandy, as did his grandson, Henry II, in 1174. After Henry II's death in 1189, his eldest surviving son, Richard the Lionheart, landed at Portsmouth as King of England.[2] However, around this time Portchester was eclipsed by the rise of the new town called Portsmouth: this may have been due to a change in tidal behaviour and sedimentation associated with the onset of the medieval warm period. Reduced flow in the Wallington River and variation in channel patterns could have induced significant changes along the Portchester shoreline, causing maritime activities to be shifted to the mouth of the harbour.[3] In 1194, King Richard I granted a charter to Portsmouth and ordered the building of a dock in the area called the Pond of the Abbess, at the mouth of the first creek on the eastern side of the harbour, which was later to become the site of the Gunwharf. Here ships could anchor, and on the creek's mudflats they could be hauled out of the water for repair or cleaning. Richard needed an alternative port to Southampton to be free of the powerful merchants and their high taxes. He needed this for his royal ships, the warships of a small fleet that travelled between England and his possessions in France.[4]

The dock was enclosed by order dated 20 May of Richard's brother, King John, in 1212:

The King to the Sheriff of Southampton. We order you, without delay, by the view of lawful men, to cause our Docks at Portsmouth to be enclosed with a Good and Strong Wall in such a manner as our beloved and faithful William, Archdeacon of Taunton will tell you, for the preservation of our Ships and Galleys: and Likewise to cause penthouses to be made to the same walls, as the same Archdeacon will also tell you, in which all our ships tackle may be safely kept, and use as much dispatch as you can in order that the same may be completed this summer, lest in the ensuing winter our ships and Galleys, and their Rigging, should incur any damage by your default; and when we know the cost it shall be accounted to you.

By implication some sort of facility already existed before William of Wrotham, Keeper and Governor of the King's Ships (and Archdeacon of Taunton), started to build his walls and the lean-to sheds to store ships' tackle and rigging. These events are seen to mark the founding of the first dockyard at Portsmouth, which now became a principal naval port, superseding the Cinque Ports.[5] We do not know what form the docks took. It is possible that a lock was built near the high water mark, and blocked with timber, brush, mud and clay walls at low tide, with a wooden breakwater, a stone wall to protect it and penthouses to store sails and ships' equipment. The lock may have been built of stone and led into a non-tidal wet dock or basin.[6]

Alternatively, there may only have been temporary mud docks with ships being dragged and docked as far up as possible on the mud at the head of the creek at high water, then closed off from the next flood tide by a wall across the creek.[7] Such mud docks may have been complemented by a tidal basin in which ships could lie afloat and also be hauled out onto a slipway, as shown in the illustration. It was from this dockyard in 1213 that King John's royal fleet of galleys joined more galleys from the Cinque Ports to achieve the first great naval victory over the French, at the Battle of the Damme.[8] A fleet led by the king sailed early in 1214 for La Rochelle and Bordeaux in an unsuccessful attempt to regain lost territory in France. Around 1228, but not for the first time, the dock was badly damaged by storms and high spring tides. It may have been this that caused Henry III in 1228 to command the Constable of Rochester 'to provide wood to fill up the basin and to make another causeway there, notwithstanding that King John had caused walls to be built close by for the protection of his vessels from storms'.[9] The dock was abandoned, and in 1253 Henry III demolished the wall and reused the stone to repair his town house.[10] Tentative evidence of the dock's location has led to the erection of a display board at the supposed site in St George's Square, Portsea. It seems that, given the paucity of sea defences at the time, the dock had not been well sited and subsequent naval docks would be built further up-harbour.

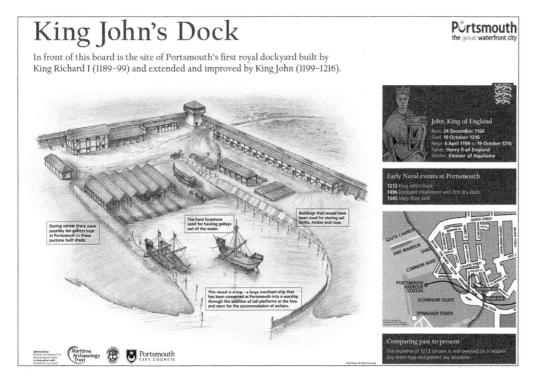

The displayboard in St George's Square, showing an artist's impression of King John's Dock in 1212. It was from here that King John's royal fleet of galleys joined more galleys from the Cinque Ports to achieve the first great naval victory over the French at the Battle of the Damme. Building of the dock had been instigated by King Richard I, who needed an alternative port to Southampton to be free of the powerful merchants and their high taxes. (Rob Kennedy)

Despite the lack of enclosed docks, Portsmouth was used to prepare expeditions to France by Henry III, and in 1346 Edward III sailed from the port with a fleet for Normandy and victory at the Battle of Crecy.[11] In 1415, King Henry V assembled his fleet at Portsmouth and Southampton, and, embarking from Portchester Castle, sailed for France and the Battle of Agincourt. On his return, he ordered the building of the Round Tower, beginning the construction of the port's defences. He also purchased land to the north of the old docks for the construction of 'The King's Dock'[12] but, following his death in 1422, it was not built and most of the king's ships were sold.

The Tudor Dockyard

The next, and highly significant, event was the ordering in 1495 by King Henry VII of what is believed to be the country's first dry dock. This was to be built on the land that Henry V had bought, in the area now occupied by No.1 Basin, and marked the founding of the current naval dockyard at Portsmouth. It was built to accommodate the *Sovereign* and *Regent*, which were bigger than their predecessors. They both drew too much water to go far up, or possibly even enter, the River Hamble, which had been used for laying up ships of Henry V's navy: this may have been one reason for the adoption of Portsmouth for the new dockyard. The designer of the dock was possibly Sir Reginald Bray, one of the trusted councillors of Henry VII, who had been made Treasurer at War and Chancellor of the Duchy of Lancaster. He was also an architect and has been credited with St George's Chapel at Windsor and Henry VII's Chapel at Westminster. In 1488, he was requested by Henry VII to dismantle the ship *Henry Grace à Dieu* and from the pieces construct a new ship to be called the *Sovereign*, having a displacement of 600 tons and carrying 141 serpentine cannon. It was this ship that was the first to use the new dock. The practice of dismantling wooden ships and building a new one from the pieces was a common practice and continued well into the eighteenth century. The task of overseeing the new dock fell to Robert Brygandyne, a yeoman of the Crown who had been appointed Clerk of the Ships (as the post once held by William of Wrotham was now known) in May 1495, as officer in charge of construction.[13]

Work on the dock began on 14 July 1495 and continued until 29 November, when it stopped for the winter. In this period the dock was dug out and the sides fixed: the sides were backed by stone and lined with wood, 158 loads of timber being used. Work started again on 2 February 1496, when the great gates were built using 113 loads of timber, which were sawn into 4,524ft of planking. These gates were then hung, being staggered in their position at the entrance to the dock and reaching across its width. The intervening space was filled with clay and shingle to form a watertight middle dam. All work was completed by 17 April 1496, and Brygandyne accounted for every payment made, the cost of construction being £193 0s 6¾d: this sum covered the wages and victuals of carpenters, sawyers, smiths, labourers, carters with their horses and a surveyor, and provision of timber, stones, clay and 'other stuff for the work'. Carpenters and smiths had to be obtained from Kent because they were not available in Portsmouth. Brygandyne also made an inventory, including smithy bellows, lanterns, caulking irons, chains, pick-axes and other items required for the operation of the dock.[14] Then came the great day on 25 May 1496 when the *Sovereign* entered the dry dock. Once in the dock, after gravity drainage at low tide, the entrance was sealed with the wooden gates and clay, and the remaining water was then pumped out using a bucket and chain pump worked by a horse-gin. It took between 120–140 men who were employed for a day and a night before the operation was complete. The majority of the men were employed on infilling with the clay and shingle. Getting the ship out of the dock on 31 January 1497 was a more lengthy procedure, as all the impacted clay and shingle had to be removed from between the great gates before they could be opened, and we are told it took twenty men twenty-four days to open them.[15] The *Sovereign* was fitted out for a chartered trading voyage to the Levant, and as soon as the she was out the *Regent* went in to be fitted out for service on the Scottish coast.[16]

Although the precise site of the dock is not known, it is generally thought to have been about 50ft astern of where HMS *Victory* lies today in No. 2 Dock. During the enlargement of the Great Ship Basin in late 1790s, the remains of an ancient dry dock were discovered in that position. However, it is possible that these remains may be from one of the seventeenth-century dry docks, although its construction would suggest otherwise. It was described in *The Illustrated History of Portsmouth* by William G. Gate as being:

A Tudor master shipwright drawing plans with his assistant. (From Mathew Baker's *Fragments of Ancient English Shipwrightry)*

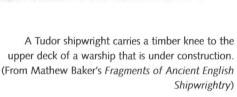

A Tudor shipwright carries a timber knee to the upper deck of a warship that is under construction. (From Mathew Baker's *Fragments of Ancient English Shipwrightry)*

formed of timber and trunnelled together, the sides being composed of whole trees. On the removal of this, many large stone cannon-balls were found. It was called Cromwell's Dock, but it seems these remains were those of the dock of 1496. It was thus described at the time of discovery: Old dock of wood, length from head of pier to head of dock, measured along the side, 330 feet on each side; the bottom of the dock 395 feet long; depth 22 feet; the wharf on the outside of the piers 40 feet on each side and depth of 22 feet.[17]

Presumably the piers were standing out from the dock sides and are where the gates were hinged from. No width of the dock is mentioned in the description, but it has been estimated to be 65ft. We are told that there were two gates, one on each side of the dock entrance hinging in opposite directions: the innermost gate hinging outwards and the outer gate hinging inwards. The length of 330ft would not have been the dock's original length, as we are told it was enlarged later in its life. The dock was a vast improvement on

anything that went before and can be seen as a turning point in the style and methods of ship repair, and the way future dockyards would be laid-out and used.[18]

In 1497, the first ships were built at the new dockyard: the *Sweepstake*, of 80 tons, which was later renamed *Katherine Pomegranate* by Henry VIII in honour of Katherine of Aragon (the pomegranate was part of the coat of arms of the city of Granada), and then the *Mary Fortune*, also of 80 tons, later renamed *Swallow*. From details of the construction of these two ships and the fitting out of *Regent* and *Sovereign*, we are able to gain an understanding of the capabilities of the new dockyard. Most of the timber was brought from the New Forest and Bere Forest and was sawn in the dockyard; on occasion iron was bought by the ton and worked up into nails, spikes, etc., at the dockyard's forge, but these items were often purchased; cordage was purchased for the manufacture (probably by seamen or shipkeepers) of the standing and running rigging; canvas was bought in for sailmaking in the yard or on the ships; everything else was purchased: deals and some cut timber from the Southampton area and other items from London, Reading, Fareham, Poole, Portsmouth itself and other places.[19]

Shipbuilding temporarily became a more important part of the dockyard's work when Henry VIII succeeded to the throne, with the construction of the *Mary Rose* and *Peter Pomegranate* helping to establish a permanent navy. Since the dry dock would have been busy with repairs to ships, the new ships may have been built on a slipway adjacent to the dock, and were constructed under the supervision of Robert Brygandyne. As well as building the two ships, for which timber, ironwork and workmanship were accounted for, Brygandyne had to fit them out and we are told that this required:

> all manner of implements and necessaries … sails, twine, marline, ropes, cables, cablets, shrouds, hawsers, buoy ropes, stays, sheets, buoy lines, tacks, lifts, top armours, streamers, standards, compasses, running glasses, tankards, bowls, dishes, lanterns, shivers of brass and pulleys, victuals and wages of men for setting up of their masts, shrouds and all other tacklings.[20]

Carpenter's tools found in the salvaged *Mary Rose*. Similar tools would have been used by house carpenters fitting out Tudor warships. (The Mary Rose Trust)

Henry VIII is often seen as the founder of the Royal Navy, since he commissioned the first ships that had an offensive role rather than being primarily transports for the army. Thus the *Mary Rose* can be considered to be the first true English warship. Her hull was of carvel construction, a recent innovation since until well into the fifteenth century English ships were clinker-built, i.e. with overlapping planks nailed together to form a skin, which was strengthened by an internal frame that could be fitted afterwards.[21] The carvel-built hull had planks laid edge-to-edge and attached by trenails to the frame (which was constructed first), giving a smooth side to the ship, making the introduction of gunports much more feasible. *Mary Rose* was part of the first generation of ships to have gunports with lids. This helped revolutionise warfare at sea – the ability to bring heavy guns lower in the hull made more layers of heavy guns possible. That this happened during the life of the *Mary Rose* is demonstrated by a change in weapons shown in the inventories for 1514 and 1540/1546, and backed by tree ring dating, which proves that extensive changes were made to her structure: probably during her rebuild at Portsmouth in 1536.[22]

The dockyard was expanded due to its strategic importance under the threat of French invasion and incursions such as that in 1545 when the *Mary Rose* sank, and the town's defences were strengthened; in 1527, 9 acres of land at 20s per acre were purchased for the dockyard.[23] During this period the Navy grew from having only twenty-one ships in 1517 to fifty-eight in 1546.[24] In 1547, the year of Henry's death, forty-one of the fifty-three ships in the Navy were based in Portsmouth.[25] Thereafter the dockyard went into relative decline, whilst those at Chatham and on the Thames prospered. The Navy contracted so that by 1578 there were only twenty-four ships – rising to thirty-four in 1588, the year of the Spanish Armada. Amongst the thirty-four warships in the fleet that fought the Armada, only three ships – *Hope*, *Nonpareil* and *Advice* – were fitted out at Portsmouth.[26] Though small, the Elizabethan Navy was very successful, and for the first time came to national prominence through the exploits of Drake, Frobisher, Grenville, Hawkins, Howard and Raleigh. They used both the queen's ships and privateers, but their forward anchorage was Plymouth. In 1623, Portsmouth's original dry dock was filled in, as was said at the time, 'to protect the dockyard from encroachments from the sea'.[27]

Good God!, what an age is this and what a world is this! that a man cannot live without playing the knave and dissimulation.

Samuel Pepys (1633–1703), *The Diary of Samuel Pepys*, 1 September 1661

The Commonwealth's Dockyard

Apart from the construction of one small ship in 1539, there was no further shipbuilding at Portsmouth since the *Peter Pomegranate* in 1510.[1] Shipbuilding resumed in 1649, under the Commonwealth, after a lapse of over a century, with the symbolically named Fourth Rate *Portsmouth*,[2] and thereafter became a more or less continuous activity for 300 years. Portsmouth had been loyal to Parliament and prospered under the Commonwealth. The construction of the *Portsmouth* was supervised by master shipwright Thomas Eastwood, who also designed her, and was followed by nine more ships in as many years, all designed by Eastwood's successor, John Tippetts. It was in 1649 that the first commissioner was appointed to take charge of the yard – William Willoughby, a colonel in the Commonwealth army. He was also Commissioner for Peace in Hampshire and arrested and imprisoned pirates, but died two years later, exhausted by his duties. He was succeeded by another army officer, Captain Robert Moulton, who also had a short period of tenure before his death in 1652, when he was replaced by Francis Willoughby, brother of William, and also an army colonel. Francis Willoughby seems to have been an able and energetic officer who soon informed his fellow commissioners on the Navy Board of the disadvantages of a dockyard, like Portsmouth, without a dry dock.

War had broken out with the Dutch in 1652, and investment in the dockyards and the Navy was at a high level. In 1656, orders were given for the construction at Portsmouth of a new double dock capable of accommodating a seventy-gun ship and a fifty-gun ship, one ahead of the other. By 1656, the yard had a slip (completed in 1651), a new ropery with two rope-walks 1,095ft long and 54ft wide, which were manned by Dutch prisoners-of-war, and a new surrounding brick wall over 400 yards in length. There were also upper and lower storehouses, upper and lower hemp houses, a tar house, block loft, office, nail loft, canvas room, hammock room, kettle room, iron loft, oil house, sail loft and houses for the rope-maker, top-maker and boat-maker and senior officials such as the master attendant. A contract was made for the construction of a wharf on either side of the building slip, which probably coincided with the construction of the entrance to the double dock. Work on the dock was delayed by a strike of shipwrights in the summer of 1657, bad weather which held up work on the foundations, the death of one of the contractors and disputes between Tippetts and other contractors, but was completed in March 1658.[3]

The dockyard was certainly busy at this time: when there were seventeen ships awaiting repairs, Willoughby wrote that, 'The multiplicity of naval affairs to be carried on here is such as scarce to leave us a minute's time from one week to another.' He listed an inventory of sixty-two anchors, 498 masts, seventy cables, 508 loads of timber, 63½ tons of hemp, 10,600 yards of made-up canvas and 7,650 yards on reels, ninety-nine barrels of tar and pitch, and 2,020 hammocks.[4] The Navy had grown rapidly to have 102 ships in 1652 when the First Dutch War started,[5] and in 1658 when Cromwell died it had 157 ships, one-third of them based at Portsmouth, which was second in importance to only Chatham.[6]

The Dockyard after the Restoration

The dockyard's growth thereafter was slow compared with Chatham, which, like Deptford and Woolwich, benefited during the Dutch Wars from its geographical position. In August 1664, a new mast pond was being dug by forty soldiers and twenty other men[7] – it was used to store the mast timbers, which were left submerged to season for twenty to thirty years; it was complemented by a new mast house in 1669.[8] It had been the custom for the dockyard commissioner to lodge with the mayor in the town of Portsmouth,

Above left: The *Royal Oak* was a 100-gun First Rate built at Portsmouth and launched in 1664 by John Tippetts, the master shipwright.

Above right: Samuel Pepys (1633–1703) was Clerk of the Acts to the Navy Board (head of the Navy Office staff) from 1660 to 1677, a post he held simultaneously with that of Secretary of the Admiralty from 1673 to 1677, before two further years in the latter post alone. (John Hayls)

but this arrangement was thought unsatisfactory by Colonel Thomas Middleton, who had been appointed as commissioner in 1664 and complained to Samuel Pepys (Clerk of the Acts to the Navy Board) about his accommodation, where, he said, 'We are forced to pack nine people in a room to sleep in, not above 16ft one way and 12ft the other. We are 26 in family, in the Mayor's house nine of which are small children. What comfort can a man have in such a condition so being together?' Pepys responded by authorising in 1665 the construction of a fine new red-brick house, with Dutch gables and surmounted by a belvedere and balcony at the top of the main staircase, giving views of the yard and the king's ships in harbour. The Commissioner's House was completed in 1666 and had stables, large gardens behind it and a large commissioner's meadow behind the adjacent terrace of five dockyard officers' houses.[9]

In March 1665, Middleton wrote to Pepys about a fire in the dockyard:

Mr St John Steventon's wife being in her time a debauched drunken woman … rose from her husband in the night, or rather towards day, went down stairs, lit a candle, and being exceedingly drunk sat in a large wicker chair which was set on fire, where she was burnt to ashes. The house escaped, which had it taken fire, would have burnt ships in dock, storehouse and what-not. The mantel of the chimney took fire, but day being at hand, people stirring, was quenched. It happened at low water, and no water in the yard considerable to put out fire if it should happen, which hope will be considered of.

In response, Pepys authorised the purchase of a fire engine at a cost of £20.[10]

In 1674, the year after the end of the Third Dutch War, the Navy Board ordered retrenchment, telling the Master Shipwright and the Clerk of the Cheque to cut the workforce to 120 shipwrights, twelve joiners, twenty caulkers, five bricklayers, one pitch heater, eight oakum boys, ten labourers, eighteen ropemakers, three sailmakers, two sawyers, ten house carpenters, one clerk of the survey and one clerk to the master shipwright; as a result, 218 men were dismissed.[11] By 1684, the Navy was in total decay and sinking at moorings unmanned, whilst the dockyards were at a standstill for want of money.

Early in November 1685, Pepys, now secretary to the Board of Admiralty again, asked Sir Anthony Deane (who had previously been master shipwright and then commissioner at Portsmouth dockyard) to propose a plan to save the Navy[12] as part of a special commission that was being set up with the agreement of the new king, James II. Deane was appointed to this post in June 1686 and the work ensued over thirty months. Three Fourth Rates were built, twenty ships rebuilt and sixty-nine repaired; the dockyards were reorganised and the stores fully replenished. A sum of £12,185 was voted by Parliament to be spent on a new dock, wharves and storehouses at Portsmouth; one dry dock was to be repaired and a further dry dock added, together with twenty new storehouses.[13] James' successor after the Glorious Revolution in 1688 was King William III, who recognised the importance of sea power and was to pursue a high level of expenditure on the Navy and the development of the dockyards, exemplified by the Historic Dockyard at Portsmouth. Deane's scheme for Portsmouth was expanded and implemented under the supervision of Edmund Dummer.

The Work of Edmund Dummer

In 1689, William embarked on war with France. Because of its Channel position, Portsmouth experienced a period of renaissance and was again second in importance to Chatham. Under the guidance of the talented Edmund Dummer (1651–1713), Surveyor of the Navy from 1692–99, a large ropery, mast pond and numerous other buildings were constructed, and two new wet docks and two dry docks were built on marshland to the north of the existing yard – some 10 acres of land being reclaimed from the sea.[14]

The first of the new dry docks, Dummer's Great Stone Dock, like those at Plymouth, was innovative – with stepped stone sides – establishing the form of dry dock that we recognise today, whereas previously dry docks had timber sides. The wide steps in the new docks, known as alters, allowed the use of shorter timbers to support ships in the dock, provided working platforms for shipwrights, reduced the volume of water required in the dock[15] and gave greater strength and water-tightness. The floor

The statue of King William III (by Van Ost) that now graces the Porter's Garden was gifted to the dockyard in 1718 by Richard Norton, a timber supplier. It depicts the king in Roman emperor style.

of each dock was still made of wooden baulks and was cambered to encourage the flow of water into a drainage channel around its edge. The Great Stone Dock (later known as the North Basin Dock, now No. 5 Dock) was ordered in 1690 and opened on 28 June 1698 with the docking of the *Royal William*. It utilised chain pumps driven by a water wheel, which was powered by water channelled from the Upper Wet Dock and discharged into the harbour. There was also a horse gin to power the pumps when the tide was too high for the water wheel to function.[16] Such limitations in the use of the water wheel may have meant that the arrangement was less than satisfactory, for it was not repeated in any other dock.

In 1691, two wet docks had been added to the original plan, and the channel leading into one of these, the Upper Wet Dock or North Basin, was converted into a second dry dock, the Lower Dry Dock (now No. 6 Dock) by the addition of an extra pair of gates in 1699. It had wooden sloping walls 20ft in height mounted on wooden piling sunk 12ft into the mud, together with a cambered floor of wood. The timber sides were replaced by stepped stone sides (alters) following a plan of 1737, and the dock was then known as the North Stone Dock.[17] To drain the dock there were two gins – large spur wheels of 17ft diameter – each moved by a team of six horses, driving cogwheels on whose shafts, which extended over the sump, were sixteen chains of buckets. The horse-gins were enclosed in a building whose ends were semicircular and whose length was almost half that of the dock.[18] Defending his work, Dummer admitted that the Lower Wet Dock and the Great Stone

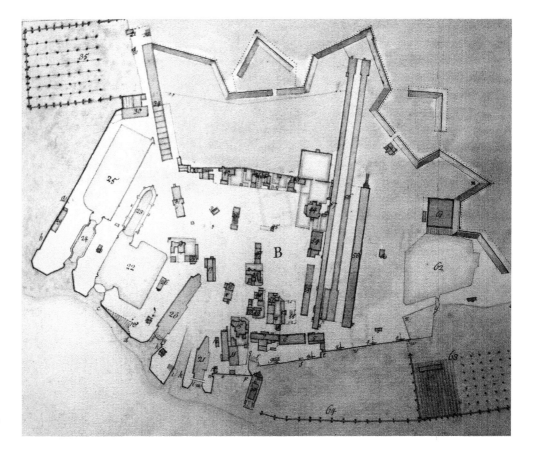

Dummer's plan of the dockyard in 1698 showed the improvements made by him since the 'Revolution' of 1688. Key: 7 Block Storehouse, 10 Broom house upon the hulk, 11 Deal Storehouse, 17 Great Storehouse, 20 Old Double Dry Dock, 21 Old Building Slip, 22 Lower Wet Dock, 23 Single Dry Dock (Great Stone Dock), 24 Lower Dry Dock, 25 Upper Wet Dock, 26 Pitch house, 30 Smiths' Shops, 34 Twelve timber boat-houses, 35 Boat Pond, 37 New Storehouse, 39 Four saw houses, 44 Commissioner's House, 47 Five saw houses, 48 Old Nail Shop, 50 Blockmakers' Shop, 51 House Carpenters' Shop, 52 New offices, 54 Tar and Yarn House, 55 Hemp Storehouse, 56 New Ropewalk including Hatchelling House and Rigging Loft, 57 Old Ropewalk, 58 Great Long Storehouse, 61 Mast Houses, 62 Old Mast Dock, 63 New Mast Dock.

Dock, 'which are to contain ships of the 1st Rate have not only wasted a long time in building and consequently large quantities of treasure', but he explained that this was due to the comparatively small rise and fall of the tide in Portsmouth harbour, which at 14ft was less than the 17ft 3in draft of a First Rate, necessitating the excavation of the docks to 5ft below the level of low water (which allowed a safety margin). This had not been necessary in the cases of the Lower Dry Dock, which accommodated ships of Third Rate downwards, or the Upper Wet Dock, which could accommodate five or six ships of Fourth Rate downwards simultaneously.[19]

In 1698, the other wet dock, the Lower Wet Dock (later known as the Great Basin), which had stone sides and double gates at its entrance, and led into the Great Stone Dock, was opened.[20] By then Portsmouth was well on the way to becoming the most important dockyard, serving a Navy that had expanded rapidly during the eight years of war: by 1697, the King's Navy contained 323 ships, including 112 with fifty guns or more. In 1698, Dummer documented a survey of the yard listing all of the buildings and docks and illustrating them with coloured plans and elevations, and even sketches of the dockyard viewed from the harbour. This gives us a very detailed picture of what the yard was like immediately after his many improvements and before the great schemes of the eighteenth century, which remodelled the yard to produce the layout and facilities still extant. There had been huge developments since 1688: in addition to the two wet docks and the two dry docks mentioned above, Dummer listed the following new facilities: guard-house, crane with double treadwheels, storehouse (rebuilt from an old one), rope-walk including hatchelling house and rigging loft, mast-dock and top-house, and offices for the Clerk of the Cheque, the Master Shipwright, the storekeepers, the Clerk of the Rope Yard and the Surveyor.[21] Dummer himself claimed:

[I]t may not be incredible (with a judicious understanding) to say that this yard (with men and materials answerable) is capable in the winter season to give ye whole Navy Royall of England a refitting whereby they may be able to put to sea in ye beginning of April and the advantage that this will be to the nation after ages (with another French war) will be able to give the best account.[22]

During his time as Surveyor of the Navy, Dummer designed and supervised the construction of the royal dockyard at Plymouth as well as designing the extension at Portsmouth. His survey of the south coast ports is a valuable and well-known historic document. He also served Arundel as Member of Parliament for approximately ten years and founded the first packet service between Falmouth, Cornwall and the West Indies. He died a bankrupt in the Fleet debtors' prison. In her account of Dummer, Celina Fox sums up his career thus:

Using elements of mathematical calculation and meticulously honed standards of empirical observation, Dummer tried to introduce a more rational, planned approach to the task of building ships and dockyards, with the help of his extraordinary draughting skills. Operating on the margins of what was technically possible, meeting with opposition from vested interests and traditional work patterns, he struggled to succeed. Today he is little recognized outside the circle of naval historians and his grandest building projects were almost wholly destroyed by later dockyard developments or bombing.[23]

The eldest son of a Hampshire gentleman farmer, he served his apprenticeship as a shipwright from 1668 under Sir John Tippetts at Portsmouth. In his 1686 account of the state of the Navy, Samuel Pepys wrote that when Dummer was apprenticed to Tippetts, he was 'mostly employed as his clerk in writing and drawing'. By 1678, Dummer was employed as an 'extra clerk' in the office of the Surveyor, who he referred to as 'my patron and friend from my youth upward'. His job was to make designs for a variety of projects – lanthorns, wet docks, lodgings at Sheerness, ships' sterns – as well as to draught ships' lines. In 1698, Dummer produced his *Survey and Description of the Principal Harbours with their Accomodations & Conveniences for Erecting Moaring [sic] Secureing and refitting the Navy Royall of England*, which gave an account of the improvements that had been made at each of the royal dockyards since 'the Revolution' of 1688, with full descriptions of the various buildings with their quantity and value, together with detailed descriptions of the new docks at Portsmouth and Plymouth.

Dummer's career as Surveyor of the Navy came to an abrupt end in December 1698, when he was suspended without warning following a dispute with John Fitch, a long-established supplier of

The *Association* was a Second Rate launched at Portsmouth on New Year's Day 1697. On the night of 22 October 1707, commanded by Captain Edmund Loades, as flagship of Sir Admiral Cloudesley Shovell (who was on board), she was returning from the Mediterranean with a twenty-one-ship squadron and entered the mouth of the English Channel. *Association* struck the Outer Gilstone Rock off the Isles of Scilly, as depicted here, and was wrecked with the loss of her entire crew of about 800 men. Her loss, with three other ships in the squadron, was due to navigational error.

New Forest timber and the main contractor undertaking building works at Portsmouth Dockyard. The origin of the dispute went back as far as 1693, when Dummer and Sir Anthony Deane had cause to doubt Fitch's working methods and his claims for payments for 'overworks'. There had been endless construction problems, caused partly because the new docks were built on unstable reclaimed land. Fitch's workmen had damaged the entrance to the Lower Wet Dock, causing its banks to slide into the channel on the spring tide, leaving the piling exposed and vulnerable. On 20 June 1695, Dummer wrote from Portsmouth to Robert Harley, a Whig politician serving on the Commission of Public Accounts, with the news that the dam that had been constructed the previous winter to shelter the work on the new docks from the sea had been breached.

The following January and February, he was writing to excuse his attendance at Parliament on account of the need to secure the great dam in the face of terrible weather. By the end of 1696, Dummer had finally ejected Fitch because of fraudulent claims for payment and terrible workmanship. Fitch brought a case for payment against the Crown, with Attorney General Richard Haddock and Dummer himself being named as defendants. The case was first heard in the Court of Exchequer on 23 November 1696 and referred to trial. On 27 April 1697, the court ordered that the matter be referred for arbitration to four referees, two appointed on behalf of the plaintiff and two on behalf of the defendants, with an umpire to determine between them if they were unable to agree. The party arrived in Portsmouth on 2 June to examine the works and advise on repairs, accompanied by Sir Christopher Wren, in his capacity as Surveyor of the King's Works. After spending a day examining the works, the party returned to London, and on 25 August their report took 'a very strict view of the nature of the said defects' confirming Dummer's condemnation of the workmanship and ordering that the piers of the Upper Wet Dock and the work on the south side of the entrance leading to the dock be taken down and rebuilt.

The Exchequer Court issued the final decree in late June or early July 1698, stating that Fitch should have been satisfied with the amounts already paid for the work done on the contract, £13,773 14s 6½d, and rejected his claims for payment on over work not included. It was also accepted that Fitch was still owed for further contract work completed before he was turned off site, valued at £8,757 1s 5d and £2,030 18s for materials. The king received £100 'for works insufficiently done' and other discounts.

Fitch then exacted his revenge by complaining to the Admiralty that Dummer had asked for bribes in return for awarding him Navy contracts, claiming specifically that Dummer had told him he would get an immediate certificate for his bill if he made him a present of £100 and helped him with the sale of timber for Plymouth. Although Dummer conceded that he had borrowed £152 from Fitch on behalf of William Wyatt, a Bursledon shipbuilder, he denied the other charges. With accusations flying back and forth, appeals and lawsuits followed and Dummer was eventually suspended from office with effect from Christmas Eve 1698. Although his name was cleared by a civil court, which awarded him damages of £364, his career at the Navy Office was over. He was allowed the title and salary for 1699, but was not reinstated and was dismissed by the Lords of the Admiralty on 10 August 1699. Dummer went on to initiate the first transatlantic mail service in 1702 and ran an iron business, but these ended in financial ruin and he died bankrupt in 1713, reputedly in Fleet prison.[24]

The Tsar's Visit

In March 1698 the Russian tsar, Peter the Great, visited the dockyard as part of a European tour which included a four-month stay in England, the first visit of a Russian tsar to the country. The main purpose of his visit was the study of English shipbuilding to help inform his project to create a modern Russian navy. He would have seen the Third Rate *Nassau* under construction at Portsmouth and, as a guest of King William III, he reviewed the fleet there, and also visited Deptford dockyard. He is said to have studied and passed a course in shipbuilding whilst in England, having already studied the subject, and worked incognito as a carpenter for four months at the Dutch East India Company's shipyard in Amsterdam. However, Peter was not satisfied with the knowledge he gained in Amsterdam because the Dutch school of shipbuilding relied heavily on practical experience rather than theoretical knowledge and calculations. The knowledge and secrets of the craft were passed down from father to son, and to use this method of building in Russia would be impossible, as Russia did not have generations of shipwrights and riggers. To learn a more structured science, Peter proceeded to England, a country known for its more precise, formalised approach to designing and building ships.[25]

When, in April 1698, Peter finally started on his homeward journey, he did so in a royal yacht based on a Sixth Rate frigate, the *Royal Transport*, newly built at Chatham and presented to him by William III. Peter lured skilled men that he met into his service: some sixty of these returned with him to Russia, to serve at the newly established Admiralty Shipyard in St Petersburg. They included Joseph Nye and Richard Cousins, both formerly of Portsmouth Dockyard: at the time of the tsar's visit, Nye had established his own small shipbuilding yard on the Isle of Wight.[26] The tsar put his newly acquired expertise in shipbuilding to good use. At Olonetsk shipyard on the River Svir on 24 March 1703, he laid the keel of the frigate *Shtandart*, the first ship in his new Baltic fleet. On the tercentenary of Peter's visit to Portsmouth Dockyard, in 1998, a plaque to commemorate the event was laid 50 yards to the west of the *Victory*.

A statue of Peter the Great at Kronstadt, the naval town and dockyard he established on Kotlin island near St Petersburg following his 1698 royal tour, which took in both Portsmouth and Deptford dockyards.

Our Navy spreads its canvas wings,

And to the French destruction brings.

Our gallant fleet so brave appears,

It makes the French-men hang their ears.

P. Brooksby, 'The Nation's Joy for a War with Monsieur, or England's Resolution to Pluck Down France', 1690

Building the Georgian Dockyard

The eighteenth century would see further expansion of the dockyard as wars with France and Spain, and also the American war, put Portsmouth in a position of pre-eminence amongst the other dockyards. At first the developments were piecemeal, but from 1761 onwards were based largely on a grand plan that produced the layout of the Georgian yard seen today in the Historic Dockyard.

A brick chapel complete with a cupola, the first chapel in any of the dockyards, was built in 1704 following complaints from dockyard men that it was too far for them to go to St Mary's at Fratton during the day.[1] In 1717, new double gates were fitted to the entrance of the stone-sided Great Basin (Dummer's Upper Wet Dock, now known as No.1 Basin). The redundant North Basin (Dummer's Upper Wet Dock) was later to be reduced in plan but deepened,[2] and became a reservoir for water. It could be quickly drained by gravity from the new dry docks built around the Great Basin. By speeding up the time to empty them, the new system increased the availability of the dry docks. Water from the reservoir was drained into the harbour, at a more leisurely rate, by horse-gins. The work started in 1771 and was completed around 1776.[3] The reservoir still lies underneath the Block Mills and the ground to the east of them.

The Georgian dockyard includes, as its oldest buildings, the Porter's Lodge (1708), adjacent to the Main Gate (now known as the Victory Gate), and Dockyard Wall (1711), the Dockyard

Officers' Houses in Long Row (1717) and the impressive Royal Naval Academy (1733). Long Row was built to replace old officers' houses that were scattered around the yard, including those of the Clerk of the Cheque, the First Assistant to the Master Shipwright, the Storekeeper, the Clerk of the Survey and the Master Caulker, which were to be taken down and the land they stood on given over to the yard, 'these buildings being very much in the way of the works of the yard'.[4] The elegant new row of nine houses is said to have been designed by John Naish, the master shipwright at the time, and to the rear of the properties were long vegetable gardens with an even older fishpond.[5] Long Row had a line of lime trees planted in front of it: this, as requested by the yard officers, who wrote to the Commissioner, 'would not only be a means to break off the weather from the houses, but a very great ornament to the building: we therefore humbly desire your honour will be pleased (as this is the reason for planting) to give the Purveyor directions to plant 36 lime trees.'[6] No. 9 Long Row was enlarged in 1832 for the Admiral Superintendent, and was later known as Mountbatten House and occupied by the Flag Officer, Portsmouth. It has an icehouse in the garden, which is thought to date from about 1840, storing ice which was brought from the Baltic each spring. In about 1990, Nos 1–8 Long Row were converted for office use.[7]

The Porter's Lodge, the oldest remaining building in the dockyard, was a domestic building where the Dockyard Porter resided. The porter had an important role, with three functions. He guarded the dockyard's boundaries and property, and marked working hours by ringing the muster bell, which was located on the side of Boathouse 5, and closing the gate against latecomers. He policed the workers to prevent excessive theft and also sold beer to the men 'to enable them the better to carry on their labour and not to distemper them'. He was also the public face of the dockyard, the daily interface between the inside and outside communities.[8] The dockyard boundary in the sixteenth century had been a hedge and ditch with gates at some points, and this had been replaced in the late seventeenth century by wooden palisades; in 1711, a new wall was built of brick.[9] In 1727, a further 22 acres were added to the dockyard from reclaimed mud land, and by 1730, Portsmouth had become the largest dockyard

Long Row was built in 1717 to house nine dockyard officers. This picture dates from c.1910: the lime trees fronting the Row have since been felled. (Portsmouth Royal Dockyard Historical Trust)

All the above had a lodging allowance of 2½d per week (which was a small contribution to the cost of their lodgings), but that allowance was not paid to the following, who were only employed casually as required:

	Per day (s.d)
Pitch Heaters	1.3
Lime Burner	1.8
Plumber	2.6
Brazier	2.6
Mason	2.6
Locksmith	2.6
Rigger	1.6
Scavel man	1.6
Labourer	1.1
Sawyer	1.6

(Source: Workmen's Wages, 1723, NMM, PFR/2/A)

in Great Britain, with 119 officers and 2,318 men.[10] Wages in the dockyards remained practically unchanged from 1690 to 1775, when a system of payment by results was introduced. The 1660 time rates were not formally superseded until 1809.

Rates of Wages in Portsmouth Dockyard in 1723

	Per day (s.d)		Per day (s.d)
Foreman of Shipwrights	3.0	House Carpenter	1.10
Quarterman	2.6	Wheelwright	1.10
Shipwright	2.1	Master Bricklayer	2.6
Foreman of Caulkers	3.0	Bricklayer	1.8
Caulker	2.1	Master Sailmaker	3.0
Master Joiner	2.6	Sailmaker	1.10
Joiner	2.0	Blockmaker	2.1
Master House Carpenter	2.6		

The early part of the eighteenth century saw battles with Franco-Spanish fleets during the War of Spanish Succession, and other wars with Spain and France ensued, culminating in the Seven Years' War with France in 1756–63. A large navy was maintained – in 1756 there were 320 ships, including 142 ships-of-the-line (First to Fourth Rates). In the light of the experience of operating the dockyard during the Seven Years' War, expansion and development was deemed necessary, leading to an ambitious plan of 1761 which took in new ground reclaimed to the north and land to the south-east, and made extensions to it a little to the east so that the Great Basin could be extended. It was necessary to remove the entrance to the Great Stone Dock, which protruded into the basin. This was achieved by shifting the dock slightly to the east of its original setting in 1769, so that although a new head and entrance were built, the central section of the dock remained in the same position. The entrance to the basin was deepened by 2ft to give easier access to large ships.[11] The enlarged basin allowed the basin to berth twelve ships-of-the-line for repairs afloat; it was later known as No. 1 Basin.

Provision was also made for new storehouses and workshops, which would all be of substantial brick construction, rather than wooden buildings with short lives and a high fire risk. The 1761 plan was largely followed through, forming the basis for much of the work at Portsmouth until the end of the century. All the existing dry docks were to be rebuilt and enlarged, and a new one was to be added.[12] By 1774, the yard had a workforce of 2,198,[13] compared with 1,080 in 1728.

Portsmouth Dockyard Manning, 14 January 1774

Officers	22	Braziers	1
Clerks	26	Locksmiths	2
Shipwrights	861	Sawyers	134
Quarter Boys	33	Sail Makers	51
Caulkers	82	Scavelmen	75
Oakum Boys	33	Masons	4
Joiners	64	Smiths	85
House Carpenters	82	Foremen	2
Wheelwrights	3	Spinners	115
Plumbers	2	Winder-uppers	9
Bricklayers	22	Labourers	14
Pitch Heaters	3	Hatchellers	18
Riggers	67	Boys	8
Rigger's Labourers	20	Blockmakers	4
Yard Labourers	293	Teams*	75

(* of one man and two horses).

Another 685 men were employed within the dockyard on ships in Ordinary (reserve) as watchmen and for maintenance.

Timber seasoning sheds were erected in the 1770s, the last, of 100ft length, being completed in 1775. It was intended that these would allow wood to be seasoned for three years in dry conditions.[14] The tidal Camber inlet in the south-west part of the yard received materials from merchant vessels that berthed there. The old Camber, which had wooden sides and required ships alongside it to sit on the mud at low tide, proved to be inadequate during the Seven Years' War and a new facility was required. Work on the new Camber, some 660ft in length, with jetties built of Portland stone, was ordered in 1773 and completed in 1785. The

No. 5 Dock was originally built by Edmund Dummer as the Great Stone Dock in 1698, but was moved to the east of its original setting in 1769 so that the Great Basin could be extended under the 1761 Plan.

three Great Storehouses, erected to the east of the Camber, could receive materials directly from its quays. In 1801–07, a caisson was introduced at the south-eastern end of the inlet to make the 220ft beyond it non-tidal. The caisson was later replaced by a fixed bridge when the Camber's usefulness had reduced because merchant ships had become too big to use it and many materials were delivered by rail.[15] On the west side of the Camber, the Sail Loft and Rigging Store were completed in 1784, and the Semaphore Tower was later added between them.[16] All were attractive Georgian buildings, if a little less grand than the Great Storehouses, with which they shared architectural features.

New Ships for Old

In the early eighteenth century, the Navy was unable to build new ships because of financial constraints. The Admiralty resorted to breaking up and 'rebuilding' existing ships. Whilst a rebuild could involve stripping off planking to facilitate replacement of rotten

The sixty-gun Fourth Rate *Centurion* (right) was launched at Portsmouth in 1732. She became famous as the flagship of Commodore Anson in his 1740–44 circumnavigation, which included the capture of the Spanish Manila treasure galleon *Nuestra Senōra de Covadonga* (left) on 20 June 1743, near Cape Espíritu Santo in the Philippines.

nearly six months of every year between 1774 and 1779 despite this being during the American War of Independence, a busy time for the yard, and the dock was done away with after the war. The South Dock was suitable only for small ships and was often empty.[17] To overcome such problems, and to cope with the continued expansion of the Navy, the construction of four new single docks and one double dock was authorised in the second half of the eighteenth century. Dry dock workload management was also helped by the introduction of the practice of sheathing ships' bottoms with copper (to prevent boring by worms, and to reduce the amount of fouling with barnacles and weeds), replacing the deal or pine sheathing that had previously been used.

Copper Sheathing of Ships

On 15 and 16 October 1759, the Navy Board issued warrants instructing Deptford yard to take delivery from a contractor of '200 plates of copper, two feet broad, four feet long, and to weigh one pound and a quarter to the foot square, with 12,000 copper nails', and to send them by 'land carriage' to Portsmouth yard, where they were to be used in lining the bottoms and sides of the main keel of His Majesty's ships the *Norfolk* and *Panther*, which were seventy-four- and sixty-four-gun ships respectively, and two months later similar instructions were given for the sixty-gun ships *Medway* and *America* at Portsmouth. The yard officers were sent instructions on how the copper was to be applied and secured, and the work was actually done in January 1760. These were amongst the first instances of partial copper sheathing in the Navy. The first ship to be copper-sheathed all over below the waterline was the frigate *Alarm*, at Woolwich in November 1761. Trials with her and several other vessels ensued, and the chief problem found was the corrosion by electrolytic action of iron bolts and fittings on the hull which were in contact with the copper, and copper alloy bolts and fittings had to be substituted for them in new ships. In January 1777, the Admiralty ordered that all new sloops and frigates were to be fixed with copper bolts below the waterline.[18] The first ship to be fully coppered at Portsmouth was probably the *Pomona*, of twenty-eight guns, in October 1778.[19] She was a new Sixth Rate, having been launched by Thomas Raymond at Northam on

frame timbers and adjustments to suit the required dimensions, in the first half of the eighteenth century it meant the complete dismantlement in dry dock of the old ship and the construction of what was, for all intents and purposes, a completely new ship, making only scant use of timber from the old ship, if at all. Between 1714 and 1744, nineteen ships-of-the-line (First to Fourth Rates) and six frigates (Fifth and Sixth Rates) were 'rebuilt' at Portsmouth, whilst no new ships-of the-line or frigates were launched between 1713 and 1731. New-builds resumed with the launching of two Fourth Rates in 1732.

Dry Dock Capacity

Dummer's docks, Nos 5 and 6, are the oldest remaining dry docks at Portsmouth. They were rebuilt in 1769 and 1737–43 respectively. The old Double Dock and South Dock were quite ineffective: in the case of the former, this was because of the difficulty of scheduling two ships' refits simultaneously, the landward end being empty

22 September 1778,[20] and would have had copper alloy fastenings below the waterline.

Because the benefits of copper-sheathing were apparent, a programme of coppering large numbers of ships already in the fleet, including ships-of-the-line, was carried out from 1779–82. A layer of thick waterproofed paper was laid on the iron-fastened hull, as insulation, before the copper was applied, but the faith of the new Comptroller, Charles Middleton, and his colleagues on the Navy Board, in this method was misplaced, for it did not solve the corrosion problem. Within two to three years, the structures of some of these ships had deteriorated to such an extent that several of them foundered at sea or were in such poor state that they had to be condemned. The seventy-four-gun *Terrible*, coppered at Portsmouth in May–June 1779, was found to be so leaky that she had to be evacuated and destroyed in September 1781. The seventy-four-gun *Ramillies*, which was the first ship-of-the-line to be fully coppered at Portsmouth (and was re-coppered there in January–February 1780), was on her passage home from the West Indies in September 1782 when she was damaged in a hurricane off the Banks of Newfoundland and became so leaky that she had to be abandoned. Just three days later, another seventy-four, *Centaur*, which had been coppered at Portsmouth in March 1780, foundered in the same hurricane in her passage home from Jamaica. The chief cause of the loss of *Ramillies* and *Centaur* was corroded iron bolts. In 1783, the seventy-four-gun *Shrewsbury*, coppered at Portsmouth in April–May 1779, was condemned at Jamaica, and the sixty-four-gun *Exeter* was condemned at Cape of Good Hope in February 1784, 'being in a very defective state'. By then the doubts about the condition of the iron bolts in all these ships (and a thorough inspection of three seventy-fours which found that their iron bolts were in a dangerous condition) had led the Admiralty to order, on 3 July 1783, a moratorium on coppering any ships except those having copper alloy fastenings. In 1786, the decision was taken to replace the iron fastenings in ships with copper alloy ones, a programme that took until 1791 to complete.[21]

The Dockyard Workload During the American War

Refitting was a complex operation which first required that the ship was stripped of most of her rigging and all of her stores and guns. From 1780 onwards, the removal of guns was often undertaken at Spithead, so that the ship was lighter for the tricky process of navigating her into Portsmouth harbour and the dock, but this was a slow operation and was prone to delays from bad weather. The docked ship's bottom was then cleaned and repaired, and the ship was then prepared for sea again. The refit of a seventy-four-gun ship typically took fourteen weeks, but only three weeks of this time was actually spent in dock, whilst five weeks were spent preparing the ship for docking and six weeks for preparing for sea again.[22] A refit involved the work of large numbers of tradesmen, especially shipwrights. For example, the ill-fated *Ramillies* was docked in the South Basin Dock on 11 November 1778 for what was a long period by the above standards, lasting twelve weeks, and in the first week 42 shipwrights were employed on her. In the ten successive weeks thereafter, the numbers of shipwrights employed on her were 61, 130, 160, 175, 79, 76, 153, 140, 99 and 70 respectively. The ship was undocked on 3 January 1779, and the work continued with the ship alongside in the basin, being completed in the first week of February 1779. From the second week of the refit, caulkers and joiners were also employed, in weeks two to eleven the numbers of caulkers being 32, 6, 7, 0, 10, 4, 32, 21, 0 and 18 respectively, and in weeks two to twelve the numbers of joiners were 2, 2, 6, 9, 9, 7, 7, 8, 9, 9 and 10 respectively. House carpenters were employed on her in only one week, and were 5 in number. During her refit, the bottom of *Ramillies* was copper-sheathed, which must have contributed to the length of time spent on the refit. Rather shorter was the four-week refit of the sixty-four-gun *Worcester* in the South Basin Dock in January 1779 (after the seventy-four-gun *Berwick* had occupied the dock vacated by *Ramillies* for two days only whilst she was being fitted for Channel service). During *Worcester*'s refit, the numbers of shipwrights employed on her were 10, 90, 213 and 70 respectively in successive weeks, whilst the numbers of caulkers was 42, 63 and 3 in weeks two to four of the refit, joiners numbered 13 and 17 in weeks three to four, and there were

2 house carpenters in week three. *Worcester* was apparently not copper-sheathed during the refit.[23]

To illustrate the docks' workload, the numbers of ships dry-docked during the American War (1774–83) were:[24]

Year	North Dock	North Basin Dock	South Basin Dock	Double Dock	South Dock	Total
1774	0	14	14	1	5	34
1775	0	27	15	4	4	50
1776	0	20	17	4	5	46
1777	17	11	14	7	2	51
1778	19	14	16	1	8	58
1779	22	17	20	3	8	70
1780	17	20	8	19	0	64
1781	3	15	6	9	0	33
1782	13	20	5	9	0	47
1783	1	59	8	4	0	72

The figures for 1783 were inflated by the number of ships docked for very short periods as they were prepared for laying up, the war having ended: this accounts for forty of the seventy-two ships docked in that year. North Dock was closed for repairs from 2 April 1775 to 25 March 1777, due to repeated troubles, some stemming from bad workmanship in hanging the gates: these problems led to the demotion of the foreman of the house carpenters 'to the station of a common working man', the reprimand and suspension for three months of the Master House-Carpenter and reprimands for the Master Shipwright and his assistants. On the instructions of the Navy Board, the Master House-Carpenter was told by the dockyard commissioner, 'that it is entirely owing to his long services, and the good character that he has hitherto borne, that their Lordships have not entirely dismissed him from his employment'.[25] North Basin Dock was closed from 22 September 1774 to 29 January 1775 due to the collapse of the basin wall because of a cavity beneath the wall's foundations 'caused by a spring or running sand'.[26] During the American War, three new Third Rates, three Sixth Rates and two sloops were launched between 1775 and 1781.[27]

New Slips and Docks

Building ships for the Navy was to remain an important activity at Portsmouth. Under the 1761 plan, the two small building slips (one just south of South Dock and one which adjoined the South Basin) were replaced by three large ones (Nos 1–3 Slips) on 14 acres of new reclaimed ground at the north end of the yard. The work was authorised in 1764 and the first two were given simple wooden roofs without heads or sides in 1776.[28] The third slip was not completed until 1784.[29] The three slips were to be fully enclosed within wooden sheds from 1814 onwards to protect ships under construction from the weather as part of a drive to overcome the problems of dry rot in ships' timbers. These structures consisted of wooden trussed roofs covered in slate with many glazed windows, supported at the sides by wooden pillars, and similar covers were built over the dry docks.[30] Between 1750 and 1840, twenty-six ships-of-the-line (First, Second and Third Rates) were built, as well as more than eighty frigates and smaller vessels. During wars, there was greater dependence on commercial yards for new construction because the dockyards were often heavily committed with refits and repairs.

Work on the first of the new dry docks, the South Basin Dock, now No. 4 Dock, began in 1767 (under the 1761 plan) on the site of the old slip on the east side of the Great Basin, and was completed in 1772. The next dock, now No. 1 Dock, was begun in 1789 and completed in 1801: it was entered from the harbour side (and was rebuilt in 1859). The South Basin Dock adjoined the Great Basin, and under a scheme devised by Samuel Bentham this basin was extended to the east, over the site of the old Double Dock, and two further docks, now Nos 2 and 3, were added. Work on these two docks began in 1799 and was completed in 1802 and 1803 respectively. Work on the new double dock (to become Nos 7 and 10 Docks) began in 1797: it was to be rebuilt in 1858. The new docks had stone sides and masonry floors. Nos 2 and 3 Docks incorporated an innovation devised by Samuel Bentham, Inspector General of Navy Works, to overcome the problems experienced with the unstable subsoil: an inverted masonry arch was built beneath the dock floor, replacing the timber floor and binding the sides and especially the gate piers together.[31] The enlarged basin was completed on 12 June 1801, when HMS *Britannia* entered it, an event commemorated by an inscription on both walls.[32] No. 5 Dock still has dock gates rather than a caisson, but the original timber gates have been replaced by metal ones.[33]

The Ropery and Fires

Two disastrous fires occurred in the ropery complex in 1760 and 1770. The 1760 fire broke out on 3 July just after midnight in the upper floor of the Hatchellers' Loft, where large quantities of dry flax and hemp were stored, and soon got out of control, spreading to a large storehouse that contained large quantities of pitch, tar, turpentine and other combustible materials, and to other buildings. It was thought that lightning caused the fire, as the night was described as 'excessively tempestuous', with a great thunderstorm raging at the time. The rain increased in intensity and continued

for some hours, which was to be the saving of the dockyard, but great damage had been done.[34] The old Hemp House, Laying House, Spinning House, Long Storehouse and the new Hemp House, with the stores within them, were all destroyed, as were the kitchen, laundry, wash house and brew house belonging to the Commissioner's House. Assistance in fighting the fire was given by 'Admiral Durell, the officers and people of the Gunwharf, the seamen, marines and soldiers from the ships at Spithead and in the Harbour, the Town, and Barracks', who helped save the Yarn House and the Rigging House.[35]

On 27 July 1770, the second fire broke out and the stores and material lost were said to be equivalent to that equipping thirty men-of-war. It was discovered at 4.30 a.m. in the Laying House and led to the destruction of the Laying and Spinning House, Long Storehouse, New Hemp House, Block Loft, House Carpenters' Shop and Mast Houses nearby. The cause was probably arson, for eleven musket cartridges were found buried in shavings in the Laying

No. 1 Dock was completed in 1801 and now holds the preserved monitor *M33*. The entrance to the dock was blocked when the Pitch House and Boat House Jetties were combined and rebuilt to form the Victory Jetty.

House.[36] John Wesley, who visited the dockyard soon afterwards, recorded his impressions:

> I walk around the dock, much larger than any in England. The late fire began in a place where no one comes, just at low water, at a time when all were fast asleep, so that no one could doubt of it being done by design. It spread with amazing violence among tow and cordage and dry wood, that none could come near without the utmost danger, nor was anything expected but that the whole dock would be consumed if not the town, but God would not permit this.[37]

Building the Great Ropehouse, some 1,030ft in length, began in 1771 to replace the burnt-out ropery, which had had separate spinning and laying houses, as well as hemp houses and tar cellars. The new building, which was completed in 1775, was known as a double ropehouse because it combined the spinning and laying houses in the same building: it remains in use in the dockyard today as a storehouse. Also remaining are the Hatchelling House and the Hemp House (both built in 1771) and the Tarring House of

The former Great Ropehouse, now No. 18 Storehouse, was built on the site of the old ropehouse (which had been destroyed by fire in 1770) and completed in 1775. It was rebuilt in 1960 with a new roof and is 1,030ft long.

1747, which survived the two fires. For expediency's sake, the new buildings were built on the site of the old ropeyard rather than on the site proposed for a new ropeyard in the 1761 plan, because the reclaimed land was not yet available. This was less than ideal, for the new site would have caused much less inconvenience to traffic in the dockyard.

To supply the ropery, the best hemp came from the Baltic ports of Riga and St Petersburg. In 1813, records show that 677 tons of hemp in store at Portsmouth came from St Petersburg, whilst 219 tons came from Riga. Lesser quantities came from the Adriatic (71 tons) and Spain (125 tons). A further 1,800 tons of hemp was on order from the contractors in St Petersburg and Riga.[38] Hemp stored in bales in the Hemp House was taken for hatchelling, which involved straightening and untangling the fibres by drawing them across rows of spikes on boards. It was then transferred via a covered bridge to the spinning lofts on the upper floors of the Ropehouse, where yarn was spun, after which it was impregnated with tar in the Tarring House to prevent rotting. This building was altered internally in 1774 to allow a horse-wheel to be inserted as a substitute for the men previously used in this disagreeable task.[39] It was recorded that:

> Lord Sandwich (First Lord of the Admiralty), in the minutes of his visitation to the dockyards, takes great pleasure in having abolished the indecent and inhuman custom, as he considered it, of naked men running round in a wheel, to draw the yarn through the tar, and for substituting horses in their stead; the next step of the progressive improvement will be that of getting rid of the horses by the adoption of the steam engine.[40]

Then, on the ground floor of the Ropehouse, the tarred yarn was twisted into rope as the forming machine travelled the length of the laying floor. The spinning and forming were manual operations, the latter being very strenuous work, it being reported that:

> The laying of a cable [a heavy rope] of twenty-three inches [diameter] is performed by the simultaneous exertion of 180 men, and requires upwards of an hour of the most strenuous exertion of strength especially on the part of those at the cranks, who not infrequently burst a blood-vessel by the severity of their labour.[41]

Although steam power was introduced at other ropeyards, it seems that its use at Portsmouth was limited to a 6hp steam engine replacing the horses in the Tarring House, that passed the yarns through heated tar and drew them through the aperture of an iron plate. This tarring machine was designed by the master ropemaker, Mr Parsons.[42] In 1813, the manning of the ropeyard totalled 347, comprising the Clerk of the Ropeyard and two extra clerks, three foremen, four layers, 151 spinners, thirty apprentices, eighty-five labourers, forty-two houseboys, one foreman of line spinners, four line and service spinners, seventeen hemp dressers and ten wheelboys. Of these, four were on sick leave and one was to be discharged, absent without leave.[43]

There was another fire on 7 December 1776, in the new Ropehouse, started by the arsonist James Aitken, aka Jack the Painter or John the Painter, who was sympathetic to the American Independence cause. He was an itinerant house painter and common felon, born in Edinburgh as the son of a blacksmith, and had spent two years in America. He first came to the Portsmouth area in January or February 1776 and found employment as a journeyman painter in the village of Titchfield, working for a painter called Golding or Goulding. He was not employed in the dockyard, but in his spare time would wander in, feigning interest in the painters' shop, though his real focus was on the ropeyard, with its wealth of flammable materials. He left Portsmouth and travelled to visit the dockyards in Plymouth, Chatham, Deptford and Woolwich, before going to Paris. In Paris he sought out Silas Deane, the new American Congress's representative there, somehow found his residence at the Hôtel d'Entragues and through sheer persistence managed to secure an interview with the rather inexperienced diplomat. Aitken explained his plans to set fire to the dockyards in England, and showed Deane a sketch of a pocket-sized incendiary canister, which could smoulder for hours before bursting into flames. It measured 10in by 3in by 3¾in, and looked like a lantern; the top was bored with air holes, and the chamber accommodated a candle, below which were combustibles such as tar, oil and matches, which would ignite when the candle burned down. Deane encouraged him in his plans and gave him a small amount of money, about £3, a passport and, according to Aitken's account, a promissory note for £300 to be paid by a contact in London. Returning to England, Aitken had the special

canister made for him by a brazier's apprentice in Canterbury. He returned to Portsmouth on 5 December 1776, and on the following day entered the dockyard without being stopped. After the workers had left for the day, he planted his canister in the Hemp House, underneath a pile of loose hemp, and in the Ropehouse placed other improvised combustibles: vials of turpentine with hemp in the mouth, alongside which gunpowder, paper and hemp was laid, and turpentine was sprinkled liberally around. Aitken attempted to set fire to them, but his matches failed him and he had to make his escape. However, night had fallen and he found himself locked in the Ropehouse; he banged on the door and eventually attracted the attention of a caretaker, Richard Faithful, and a man who repaired the ropemakers' tools, Richard Voke, who let him out – something that they later lost their jobs for.

The next day, having obtained some more matches, he re-entered the yard in the morning and, after the ropemakers had left for the day, started three fires in the Ropehouse, and tried unsuccessfully to light another in the Hemp House. At 4.30 p.m., the caretaker Richard Faithful was chatting upstairs to Richard Voke when he smelled smoke. Faithful battled through the smoke to the ground floor and made his escape from the burning building. Voke went to rescue his tools, but the smoke was now too thick for him to carry them downstairs. He scurried to the nearest window, which he straddled whilst onlookers gathered below and 'pulled off their clothes for him to fall upon'. However, a ladder was found and Voke scrambled down to safety. Meanwhile, Aitken headed out of town, running most of the way before hitching a lift on a cart, and made his way to London. He then went to Bristol, where he started more fires, but the authorities were on his trail and he was eventually apprehended in Odiham, Hampshire.[44]

After trial at Winchester, Aitken was hanged on 64ft-high gallows erected on the mast of the *Arethusa* at the dockyard gate on 10 March 1777,[45] and his remains were suspended in chains at Fort Blockhouse for several years as a warning to others. The damage in the dockyard had been extensive, despite the efforts of 'a multitude endeavouring to extinguish the fire'.[46] Both the Spinning and the Laying Houses had been damaged, as had a storehouse, the new Hemp House, the Block Loft, the House Carpenters' shop and a nearby masthouse, together with its masts. Large quantities of hemp, tar and cables were lost, as well as the carpenters' and boatswains' stores and the fitted rigging of twenty-three ships in Ordinary. The final damage was estimated at £49,879 3s 4¼d.[47] The Commissioner, James Gambier, was so disturbed by this incident, which he said he had investigated with the 'utmost precision and diligence', that he took to his bed, complaining to the Navy Board, who had been critical of a letter he had sent, 'I cannot suffer my Conduct in any shape whatever, to be arraigned after my unremitting fatigue and trouble.' When ordered shortly afterwards to allow no strangers into the yard, he was stung by the implied rebuke and replied that he had 'too sensibly suffered both in his constitution and pocket, not to most attentively explore every means in human power to obviate and prevent a repetition of so fatal and melancholy an event'.[48]

The ropemakers were sent to Woolwich, Chatham and Plymouth until production could be resumed at Portsmouth.[49] Because they were of more substantial construction than the previous buildings the new buildings survived, but their interiors had to be rebuilt, and within a year they were complete.[50] In 1805, Portsmouth produced 1,633 tons of rope, against 2,120 tons at Chatham, 1,938 tons at Plymouth and 7,696 tons obtained from contractors.[51] The ropemakers had a ceremonial role whenever the sovereign paid a state visit to Portsmouth. It was an ancient custom for them to precede the royal carriage from the boundary of the borough, uniformly dressed, bearing white staves and the national flag, and wearing blue sashes across the shoulder. Such a ceremony took place in 1814 on the visit of the Regent to the town.[52]

Chatham became the first dockyard ropery to have a steam engine powering its main machinery, but Portsmouth's was never similarly equipped.[53] With the decline in the sailing Navy, demand for rope reduced and Portsmouth's ropery ceased production in 1868 (whilst rope production continued at Devonport and Chatham). In that year, seventy-six ropemakers and spinners were discharged at Portsmouth, with pensions or gratuities, amongst a total of 2,000 discharges that resulted from 'reconstruction of the navy by the substitution of ironclad ships'.[54] The buildings were then used as storehouses and their interiors modernised. So effectively had the Ropehouse split the dockyard in two that the opportunity was taken to drive an archway through it. In 1960, the Ropehouse (by then known as No. 18 Storehouse) was reconstructed with a new roof (minus the original dormer windows) and new entrances.

The archway built into the former Great Ropehouse after its conversion into a storehouse, to allow traffic to follow a shorter route in this part of the dockyard.

The Hatchelling House and Hemp House (1771) became No. 19 Storehouse, whilst the Hemp Tarring House (1771) became a boiler shop and, later, Scott Road Offices. Running parallel to the Ropehouse, on its south side, is Anchor Lane, so called because it was used to store anchors, gantry cranes being fitted above to lift the anchors – though no trace of these remains.[55]

Other Georgian Buildings

Other buildings still surviving from the second half of the eighteenth century include the three Great Storehouses (1763, 1776 and 1782), which can be seen on the left of the road between the Victory Gate and HMS *Victory*. Built of brick with Portland stone detailing, these are the most architecturally distinguished of all the surviving dockyard storehouses. Parts of their interior structures were built of reused timbers taken from ships or dockyard buildings that were being dismantled. The construction of the second two stores was considerably delayed by the fires in the dockyard in 1760, 1770 and 1776. They originally stood on the water's edge and there was a jetty on the harbour side of the buildings, allowing boats to come and unload and collect materials, gear and stores, often for transfer to ships lying at Spithead, since sailing ships would usually avoid the tricky business of getting in and out of the harbour unless they were going into dock for repairs. There were manually operated cranes to lift goods between the boats and jetty, and building-mounted winches that lifted them into or out of the storehouses.[56] No. 11

Storehouse, built in 1763 and previously known as the Present Use Storehouse, the oldest and most northerly of the three, now houses the National Museum of the Royal Navy, including the Sailing Navy Gallery and the Nelson Exhibition. No. 10 Storehouse, built in 1776, was previously known as the Middle or Clock Store, and included an archway in the middle of the ground floor for taking in cordage from the Ropehouse, which was situated on the road directly opposite the store. It now provides accommodation for the Royal Naval Museum's 20th Century Gallery.

The clockhouse now in No. 10 Storehouse was installed in 1992, replacing the original which had been destroyed by German incendiary bombs during the 1941 blitz on Portsmouth. The clock on the replica tower came from Bristol Grammar School.[57] No. 9 Storehouse was built in 1782 and was previously known as the South Store: it now houses retail premises and storerooms for Portsmouth Naval Base Property Trust and Portsmouth Royal Dockyard Historical Trust. In 1813, the three Great Storehouses housed a huge range of stores, as illustrated in the table showing the types of stores held:[58]

No. 9 Storehouse was completed in 1782, the last of the three great storehouses to be built. On the right can be seen the figurehead of HMS *Benbow* (1813) and the clock tower of No. 10 Storehouse.

No. 11 Storehouse is the oldest of the three great storehouses, having been completed in 1763 as the Present Use Storehouse. It is now occupied by the National Museum of the Royal Navy.

Table: Contents of the Great Storehouses, 1813.

	Present-use Storehouse	Middle Storehouse	South Storehouse
Dimensions	186ft x 45ft in the clear	204ft x 45ft in the clear	186ft x 45ft 9in in the clear
Attic	Mast hoops, thrums, sheathing paper, office records, taylor's room, oil in bottles, new canvas, old buntin	Old canvas for boatswains and riggers and lots of old canvas	Appropriated sails, signal balls
2nd Floor	Colours, flags, buntin, women's workroom, office records, canvas, fearnought and kersey	Spare sails, hammocks	Appropriated sails, cordage of 3 to 1½in
1st Floor	Stops, marine clothing and beds, compasses, glass lamps, brass room, stops	Cordage from 4½ to ¾in. Boltrope, Hawsers laid from 9 to 5in, and cables from 6½ to 6in	Appropriated sails, remanufactured cordage. Fitting sailroom
Ground Floor	Iron nails, locks, hinges, lanthorns, fishing gear, patterns, candles, hand and deep sea leads, leather, lead scuppers and pipe, copper nails, pump gear, brass shivers, bells, chains, rotherboat	Cables 22 and 25in	Cables 23 and 24in
		Hawsers and cables 5½ to 2in	Cables from 16½ to 14in. Rope for harbour service. Cables from 21 to 17½in
Cellar	13 arches containing tallow, oil of sorts, old lead, black varnish and vinegar	12 arches containing tar and pitch	13 arches containing pitch, old iron, new buoys, turpentine, rosin, Russia bar copper, charcoal and iron buoys

Notes: boltrope – rope sewn into the edge of a sail to strengthen it; fearnought – woollen cloth used in seamen's clothing; kersey – coarse woollen cloth used for making colours and for lining scarf joints; lanthorn – an old name for lanterns; rother – rudder; shiver – a pulley used in rigging blocks; thrum – a short piece of waste thread or yarn.

Built around the same time as the Great Storehouses, and opposite them, running parallel with the Great Ropehouse, are three more Georgian storehouses – the East Sea Store (1771, now No. 15 Storehouse), and the West and East Hemp Houses (1771 and 1782, now Nos 16 and 17 Storehouses respectively). The West Sea Store (1771) was destroyed in 1941. Another group of four Georgian storehouses was constructed to the east of the Great Basin between 1786 and 1790, of which three survive – the twin South-East and South-West Buildings and the North-West Building. The South-East (now No. 25 Store) is architecturally the most attractive, with pedimented north and south elevations and having been altered less than the other two. The upper floor of the South-West building once served as a mould loft. The North-East building was destroyed in the Second World War.[59]

In Portsmouth, the Commissioner, Sir Henry Martin (who was a personal friend of the royal family), had been given permission

The South-West Storehouse, one of a group of four large storehouses built to the east of the Great Basin in the 1780s, one of which was destroyed in the Second World War.

Short Row, designed and built by Thomas Telford, was completed in 1787 to provide houses for five dockyard officers. Actors in period dress pose at the 2005 International Festival of the Sea. (Portsmouth Royal Dockyard Historical Trust)

to construct a new residence that would befit his station and be elegant enough to act as a home for the king and the royal family when they visited Portsmouth. The new residence was designed by Samuel Wyatt, assisted by his brother Joseph, is generally thought to have been commenced in 1784 and was completed on 10 March 1786. At this time Portsmouth dockyard, along with other royal dockyards, was in the process of major rebuilding. The Commissioner complained that the Admiralty Surveyor of Works for Royal Dockyards (John Marquand) could not devote enough time to oversee the building of the new residence. Samuel Wyatt persuaded Sir Henry Martin to have Thomas Telford appointed as Clerk of Works for Portsmouth Dockyard, where he would be in constant supervision of the new building. Telford (1757–1834) was the son of a shepherd and was born in Westerkirk, Scotland. At the age of 14 he was apprenticed to a stonemason and worked for a time in Edinburgh, before moving in 1782 to London, where he worked as a stonemason under Samuel Wyatt, who was responsible

for the building work on Somerset House. During Telford's two-year stay in Portsmouth, from 1784–86, he also undertook other work in the dockyard, such as acting as Clerk of Works for St Ann's Church (1785), which was built in eight months to the design of John Marquand, the Admiralty Surveyor, and the builders were Thomas Parlby & Son. It was apparently capable of accommodating 2,000 worshippers and survives today, although the east end of the church suffered bomb damage in 1941 and rebuilding shortened it by two bays. Short Row (1787), a terrace of five houses built for dockyard officers including the surgeon, the clerk of the ropeyard, the master ropemaker and the boatswain,[60] was designed and supervised by Thomas Telford in his first architectural commission.[61] It supplemented the earlier Long Row, which housed nine other senior dockyard officers.

The Commissioner's House (later Admiralty House) was built on a site just to the north of the Academy. It replaced the seventeenth-century Commissioner's House and was the grandest of the dockyard commissioner's houses on account of the need to provide suitable accommodation during the visits to it of King George III and his third

St Ann's Church was built in 1785 to replace the church demolished to make way for a new Commissioner's House. St Ann's was heavily damaged in 1941 bombing, and was reduced in size during the rebuilding.

son, Prince William Henry (a lieutenant RN, who was attracted to the commissioner's daughter and had visited several times in 1785).[62] The king reviewed the fleet at Spithead in 1773 and 1778, and in 1794 he went aboard the flagship *Queen Charlotte* at Spithead after Howe's victory over the French on the Glorious First of June. The old Commissioner's House had been deemed to be old and inconvenient, and occupied (with the large gardens attached) a site much required for other purposes. In order to make the site available, the then existing dockyard church and the Powder Magazine had to be demolished, and the new St Ann's Church was built to replace the old church. When the Navy Board was dissolved in 1832, the Commissioner's House became the official residence of the Port Admiral. At the same time, No. 9 Long Row was enlarged and appropriated for the Admiral Superintendent of the dockyard.[63]

Office work was accommodated in the South Office Block (1786–89) which now faces HMS *Victory*. It was purpose-built and had corridors on each floor for internal communication, a feature lacking in earlier dockyard office buildings. However, many of the ground-floor offices still retained their own external doors, perhaps for reasons of status. An adjacent storehouse, to a similar exterior design on the east side of the first office block, was later altered into office accommodation and was linked to the first building by an impressive archway block.[64] The figurehead over the porch of the South Office Block is from the royal yacht *Victoria and Albert*, which was launched in 1843 at Pembroke Dockyard and broken up at Portsmouth Dockyard in 1868.[65]

Bentham's Improvements

In March 1796, Sir Samuel Bentham (1757–1831) was appointed Inspector General of Navy Works by the Admiralty and started to modernise the dockyards. The son of an attorney, he had been apprenticed at the age of 14 to a shipwright at Woolwich Dockyard, serving there and at Chatham Dockyard, before completing his training at the Royal Naval Academy in Portsmouth, where he spent the last two years of his seven years' apprenticeship. In 1780 he moved to Russia, where was employed in the service of Prince Potemkin, who had an establishment designed to promote the introduction of various arts of civilisation. Initially hired as

The South Office Block (1786–89) has a figurehead from the royal yacht *Victoria and Albert* (1843) over the porch.

a shipbuilder, he was commissioned into the army and soon discovered other opportunities to use his talents as an engineer and inventor, constructing industrial machinery and experimenting with steel production. He helped establish manufacturing industries including sailcloth and rope-making as part of a drive to modernise the Russian navy, and designed and constructed many novel inventions, including inventing a number of woodworking machines (which he patented in England), an amphibious vessel and an articulated barge built for Empress Catherine the Great. He gained a knighthood and rose to the rank of Brigadier General in Catherine's service. He eventually came to have complete responsibility for Potemkin's factories and workshops, and it was while considering the difficulties of supervising the large workforce that he devised the principle of central inspection and designed the first Panopticon building that would embody that principle and was later popularised by his brother, the philosopher Jeremy Bentham.

Samuel returned to England in 1791, and in 1795 the Lords Commissioners of the Admiralty asked him to design six new sailing ships with 'partitions contributing to strength, and securing the ship against foundering, as practised by the Chinese of the present day'. These were built by the shipyard of Hobbs & Hellyer at Redbridge, Hampshire, and incorporated a number of other novel features such as interchangeable parts for masts and spars, allowing easy maintenance while at sea. In his new post he produced a great many suggestions for improvements in the dockyards, which included the introduction of steam power and the mechanisation of many production processes. However, there was often friction with his superiors at the Navy Board, some of whom were less open to change and resented the imposition of Bentham by the Admiralty in what were traditionally Navy Board matters, and many of his suggestions were not implemented.[66]

Sir Samuel Bentham, appointed as the first head of the Navy Works department in 1796, was responsible for the modernisation of the dockyard through a series of innovations and remained in post until 1812.

In 1798, Bentham introduced the first caisson to a British dockyard when he replaced the gates which opened into the Great Basin with a boat-shaped floodable chamber which fitted into the stone grooves in the basin entrance.[67] The great advantage was its width, which allowed wagons and carriages to cross over it, obviating the need for a considerable detour.[68] In 1799, as noted earlier, Bentham came up with his innovative design for the foundations of two new dry docks at Portsmouth, using inverted masonry arches, and, under his direction, a significant technological change came to Portsmouth dockyard that year when a small steam engine (of 12hp) was installed to replace one of the two horse-gins which powered pumps to drain the Reservoir's sump into the harbour. This steam engine, designed by James Sadler, the chemist on Bentham's staff, was the first use of steam power in a Royal Dockyard and indeed the first use in the Royal Navy. Bentham also proposed that water could be pumped into the Great Basin to raise the level, giving more water over the cills for large ships entering or leaving the dry docks.[69] This overcame the great previous limitations of the old pumping systems which had meant that the depth of water in the docks was largely determined by the rise and fall of the tide, and there was a high dependence on the high spring tides, which occurred only once a fortnight, for docking and undocking.[70] It also had the potential to allow ships to be docked without being cleared of their guns and stores, though this apparently did not happen very often.[71]

A second steam engine, of 30hp, by Boulton & Watt, was ordered in 1800 to guard against mechanical failure and allow the removal of the other horse pump. Both engines and their respective boilers were installed in new brick engine houses which soon became incorporated into the larger Wood Mills complex that Bentham designed. He also organised the design and drilling of a new fresh-water well which pumped water from a depth of 274ft below ground level. The water was raised by well pumps driven by the new steam engines, and the supply was connected to a network of cast iron pipes that distributed it to different parts of the dockyard, including the provision of water cocks for fire-fighting. All this was a great advance: previously, fresh water was drawn from a number of small wells in the dockyard, but these were not conveniently situated for the wharves, and warships normally had to cross the harbour to Gosport and take on water from the Weevil Lane well, which also served the Navy's brewery which had been built in 1782.[72] The Sadler engine was a house-built table engine, and was replaced in 1807 by another, more powerful table engine made by Fenton, Murray and Wood of Leeds and, in turn, in 1830 by a Maudslay beam engine. The Boulton and Watt beam engine was replaced in 1837 by an engine made by James Watt and Co.

Bentham successfully developed the first bucket-ladder steam dredger, which was completed at Portsmouth by October 1802. Its purpose was to clear shoals in Portsmouth harbour so that large ships could enter more easily, and it could raise shingle from 14ft at a rate of 80 tons an hour. The Portsmouth officers noted in 1805 that whereas scavelmen would have taken five months to clear the yard Camber of mud, the steam dredger could perform the task in sixteen and a half days.[73]

Another of Bentham's imaginative schemes was to build over the Reservoir, thus creating a new space which he estimated at 35,000sq.ft in what was otherwise a very cramped site. A lower tier of brick vaults would cover the sump, whilst the upper tier would provide space for the lay-aside of ship's stores and the installation of Bentham's woodworking machines, which would be powered by the two new steam engines via a system of drive belts and shafts. The work on the vaulting was approved in 1800 and completed in 1805. Two parallel three-storey brick buildings were erected in 1802–03 for the Wood Mills, and between them was a secure fenced-in yard for the storage of timber. The existing engine and boiler houses were incorporated into the southern building. Above the second floor of each building were situated water tanks to provide a head of water for the new water main. The woodworking machines were predominantly saws of various types (both reciprocating and circular) which, as they were developed, progressively mechanised what was one of the most arduous tasks in the dockyard.[74]

At the same time, Bentham and his mechanist, Simon Goodrich, designed and built a metal mill to the north-east of the Wood Mills. This facility was to recycle copper sheathing removed from ships' hulls, and apparently included a copper-smelting furnace, a rolling mill and tilt hammers; by 1804, the project had been expanded to include similar equipment for the reprocessing of iron, and the two facilities were powered by a 56hp steam engine. The Metal Mills building does not survive.[75] Trials with the copper mill showed that two men working one pair of rolls with one furnaceman, one man pickling and shearing and three boys assisting could produce 6–7 tons of recycled copper a week.[76] By 1807, Portsmouth was producing 800 tons of recycled copper annually, which Goodrich stated was about two-thirds of the Navy's requirements, and by 1808 output had risen to 1,000 tons.[77] By 1813, the savings made by the Metal Mills amounted to about £41,000 a year. Bentham also established a millwrights' shop, making machine tools. According to Bentham, in 1813 the Millwrights Department, Wood Mills and Metal Mills at Portsmouth were performing work worth over a million pounds in labour and materials.[78]

The Block Mills buildings were built in 1802 to house the steam-powered block-making machines designed by Marc Isambard Brunel, and the Wood Mills. Many of the block-making machines stayed in use for over 150 years. Beneath the Block Mills is a reservoir, formerly Dummer's Upper Wet Dock, which took water drained from the dry docks around the Great Basin (now No. 1 Basin).

The Block Mills

These developments were paralleled by the design of Marc Isambard Brunel for steam-powered block-making machines. Brunel, a farmer's son, was born in 1769 in the hamlet of Hacqueville in northern France, and served in the French Navy for six years. As a royalist, he had to flee to the United States during the French Revolution, and spent several years there as an architect and engineer. He came to England in 1799 to show his method for making ships' blocks by mechanical means to the Board of Admiralty. He had collaborated with Henry Maudslay, a leading mechanical engineer and machine toolmaker, to produce working models of the proposed machines (which he was able to demonstrate to the Admiralty) and one full-size machine for making blocks of up to 9in diameter (which he had shown to Bentham). The plans were accepted and Brunel was placed in charge of installing the machines at Portsmouth Dockyard. His son, Isambard Kingdom Brunel, was born in Portsmouth in 1806. The block-making machines are often said to have been the world's first powered mass production machines. However, they were part of a more gradual progression towards fully automated machine tools, and Brunel's genius was to bring together both existing knowledge and new innovatory features of his own design in what must be regarded as a major technological leap forward in

the process of mass production. In this he was the leading figure, though almost certainly aided by Henry Maudslay, Simon Goodrich (mechanist to the Navy Board) and Samuel Bentham.[79]

The first set of machines, for medium-sized blocks (of 7–10in length), was installed in January 1803, the second set for smaller blocks (4–7in) in May 1803 and the third set for large blocks (10–18in) in March 1805. In all there were forty-five machines of twenty-two different kinds, all made by Maudslay; they included Bentham-designed saws for the preliminary cutting of the timber blanks, whilst the machines for shaping and finishing the blocks were all designed by Brunel. There were numerous changes of layout and some modification of the plant until, in September 1807, the plant was felt able to fulfil all the needs of the Navy: in 1808, 130,000 blocks were produced. Many of the machines stayed in use for over 150 years, an astonishing testament to the quality of Maudslay's workmanship.[80] The block-making building infilled the storage space between the two woodmill ranges, and the whole group soon became known as the Block Mills. Although block-making machines were to take over some of the original woodmills space, the other woodworking machines continued to produce a wide range of items including, in 1804, *inter alia*, oak hammock racks, oak tracing battens, grating battens, table legs and frames, ladders and cants cut out and rebated for bulkheads. Doubtless there were many other products, for in 1854 the output was said to include items as diverse as capstans of all sizes, steering wheels for ships of the line, handles for pumps, bells and winches, and batons for the dockyard constables.[81]

The new equipment reportedly allowed 110 skilled blockmakers – who were mostly employed by two private firms at Southampton and Plymouth – to be replaced by only ten machinists in the Block Mills.[82] Blocks were simple wood-enclosed pulleys used extensively on sailing ships. A seventy-four-gun ship required some 1,400 pulley blocks for the running gear of the sails and for handling the guns. Being used in such high quantities, they were ideal candidates for mass production methods. A pulley-block has four parts: the shell (made of elm wood and scored on its external surface to hold a rope strap), the sheave (or pulley-wheel, of lignum vitae), the iron pin for locating the latter in the shell and a metal bush, or coak, made of bell-metal and inserted into the sheave to save wear between it and the pin. All the wooden parts were made in

A mortising machine, formerly in the Block Mills, is now on display in the Dockyard Apprentice Exhibition in No. 7 Boathouse. This machine took blocks from the boring machine and enlarged the large bored hole into a slot to take the sheave.

the Block Mills, whilst the metal parts were outsourced at first but were later made in the Metal Mills. Blocks varied in size and in the number of sheaves. There were seven stages in the production of a shell, each of the first six being carried out by a different machine, in a rudimentary flow-line layout, while the final stage involved hand-finishing with a spokeshave. Manufacture of the sheave involved six different machine operations and two hand operations, including fitting the coak. Finally, the shell, sheaves and pin were assembled by hand to complete the block. By September 1807, Brunel was able to report to Goodrich that the Block Mills could, from the following month, produce all the blocks necessary for the Navy, at a saving of £17,000 a year.[83]

Annual production in Portsmouth's block mills peaked at 140,000 blocks during the Crimean War. As the Navy changed from sail to steam, the demand for blocks declined rapidly, but they continued to be produced on a limited scale until 1965, when the Block Mills closed. Examples of the original block-making machines can be seen at the Dockyard Apprentice exhibition in No. 7 Boathouse at Portsmouth, whilst others are at the Science Museum in London. The empty Block Mills buildings can still be seen on the north side of No. 1 Basin (though they are without their original chimneys: electric motors replaced the steam engines in 1911.) The Reservoir is still in occasional use, the pumps now also being electric.

Work in the central range of the Block Mills, c.1910. Overhead drive belts which were powered by steam engines in the south range can be seen. The row of machines on the left is arranged in production sequence: the boring machine has a pile of solid blocks of wood in front of it, and beyond are mortising, cornering and shaping machines.

Brunel was also instrumental in the development of large circular saw blades of up to 9ft in diameter which would have been able to tackle the largest tree trunks, which had previously been cut by the pit-sawyers.[84] By 1813, additional saws in the Wood Mills permitted cutting at the rate of 9,000–10,000ft of timber, of various sorts including 4in and 5in planks, each week.[85] Even with all these innovations, the dockyard labour force continued to increase during the Napoleonic Wars, reaching 4,300 by 1813, and at this time the dockyard was the world's largest industrial complex.

The steam engines, woodworking machines and the Block Mills attracted an enormous amount of interest and there were many visitors, both prominent dignitaries and the general public. In an event which represented a remarkable collision of epochs, Admiral Lord Nelson visited the Block Mills on the morning of 14 September 1805, the day he embarked from Portsmouth to join HMS *Victory* in St Helens Roads and thence to the Battle of Trafalgar. The hero of the sailing Navy thus glimpsed the dawn of the steam age in the Navy. Even during the time of the Napoleonic Wars, until 1815, there was a stream of foreign dignitaries and military men wishing to visit. In 1831, Princess Victoria, then aged 12, visited the dockyard (with the Duchess of Kent) as part of her education, and was shown around the Block Mills by Simon Goodrich, who as mechanist (an engineer) under Bentham had had a big involvement in their development and operation.

Bentham's achievements in modernising the dockyards were certainly impressive. Some idea of the scope of his schemes and ideas was given after his post was abolished in 1812 when he provided a list of projects he claimed to have contributed. His *Statement of Services relative to the Improvement of Manufactures requisite in Naval Arsenals* proceeded under nine headings: steam engines, sawmills, system of machinery, new tools, system of manufacturing establishments, millwrights shop, wood mills, metal mills, ropery and sailcloth manufactory. His *Statement of Services relative to the Improvement and Formation of Naval Arsenals* had twenty-one headings: accommodations for fitting out and storing, shallow docks, increased use to works, floating dams (caissons), covered docks, timber seasoning houses, wells, water works and fire extinguishing, fire proof buildings, lamps, mortar mill, roman cement mill, digging engine, mud barges, moveable steam engines, dams, foundation masses, repair of old works, saving of space, new naval arsenal,

desiderata in an arsenal. He also developed principles for the management of industrial organisations which resonated with the new processes of mass production and the division of labour that he was introducing. They included ideas of individual responsibility – each individual to act on his own conscience in the public interest, the classification of labour by intelligence and skills, mass education to facilitate mobility between classes (and motivation of the workforce by the desire to rise in class) and to optimise worker performance, and central financial control through the use of accountancy. He supported his claims, which were intended as a vindication of his role, with marginal references to the dates of official letters.[86]

The French Revolutionary and Napoleonic Wars

In the period of the Revolutionary and Napoleonic Wars, 1793–1815, Portsmouth carried out more than a third of the fleet's refits. This was its primary role; major upgrades, rebuilds and conversions were a secondary priority, whilst new construction was third in importance. The workload of the yard (measured by ton dock days, calculated as the sum of the number of dock days for each vessel multiplied by the vessel's tonnage) effectively doubled during this period. This was due partly to the progressive aging of the fleet, which required more work to keep it in a seaworthy condition, and also because of the increase in the size of the fleet – which grew from 390 ships of all sizes in 1793 to a maximum of 979 in 1809.[87] Handling the increased workload was helped, of course, by the additional capacity provided by the new dry docks and workshop facilities added to the yard during this period.

Between 1793 and 1815, 1,810 ships passed through the dock complex for refitting, fitting or repair.[88] There was comparatively little new construction during the wars. Only one ship was laid down between 1793 and 1803, this being the Second Rate *Dreadnought*, ordered in 1797 and launched in 1801,[89] though one sloop and one Second Rate were under construction when war broke out and were launched in October 1793 and June 1794 respectively.[90] Between 1804 and 1815 there was rather more new construction activity, with three ships-of-the-line, four frigates (Fifth and Sixth Rates) and six smaller ships launched.[91]

he exact Representation of Launching the Prince of Wales Man of War, before their Majesties at Portsmouth.

The workforce expanded by 92 per cent between 1793 and 1813,[92] compared with a ton dock day increase of 107 per cent. This suggests improved productivity, despite the fact that there had been a huge increase in the number of apprentices (who might be assumed to be less productive than other workers). The productivity gain was probably more than these figures suggest, because certain activities (such as blockmaking and the recycling of copper) had by 1813 been brought in-house rather than being outsourced. Taking this into account, the productivity increase has been estimated at 27 per cent,[93] some of which must have been attributable to Bentham's reforms.

The launch of the seventy-four-gun Second Rate *Prince of Wales* at Portsmouth in 1794. After active service in the French wars, she was broken up at Portsmouth in 1822.

Portsmouth Dockyard Establishment, September 1813 (Source: TNA ADM 7/593)

Commissioner	1
Clerks to	3
Master Attendants	3
Clerks to	3
Master Shipwright	1
Assistants to	3
Clerks to	5
Clerk of the Cheque	1
Clerks to	15
Temporary clerks to	2
Clerk of the Survey	1
Clerks to	9
Temporary clerks to	8
Chaplain	1
Surgeon	1
Assistant Surgeon	1
Master Measurer	1
Submeasurers	22
Converters	2
Master Mastmaker	1
Master Boatbuilder	1
Master Joiner	1
Master House Carpenter	1
Master Smith	1
Master Bricklayer	1
Master Painter	1
Master Sailmaker	1
Master Rigger	1
Master Boatswain	1
Warder	1
Shipwrights, new work afloat and in the yard	
Foremen	6
Quartermen	44
Cabin Keepers	9
Single Station	81
Shipwrights	721
Apprentices	269
Shipwrights, Mast House	
Foreman	1
Quartermen	4
Shipwrights	87

Apprentices	13
Carpenters	9
Shipwrights, Boat House	
Foreman	1
Quartermen	3
Shipwrights	104
Apprentices	18
Shipwrights, Capstan House	
Quarterman	1
Shipwrights	19
Shipwrights, Tophouse	
Quarterman	1
Shipwrights	19
Apprentices	2
Squaremakers	6
Cooper	1
Blockmakers	6
Apprentice	1
Foreman of Caulkers	1
Quartermen	9
Cabin Keeper	1
Caulke Apprentices	16
Pitch Heaters	2
Oakam Boys	44
Foreman of Joiners	1
Cabin Keeper	1
Joiners	143
Apprentices	17
Foreman of Smiths	1
Smiths	143
Apprentices	32
Foreman of House Carpenters	1
Cabin Keeper	1
House Carpenters	228
Apprentices	28
Wheelwrights	3
Plumbers	9
Apprentice	1
Brazier	1
Apprentice	1
Founders	2
Tinman	1

Foreman of Bricklayers	1
Cabin Keeper	1
Bricklayers	34
Apprentices	2
Labourers	42
Foreman of Sailmakers	1
Cabin Keeper	1
Sailmakers	58
Apprentices	14
Foreman of Masons	1
Masons Apprentices	3
Foreman of Painters	1
Cabin Keeper	1
Painter	35
Apprentices	7
Colour Grinders	2
Labourers	13
Sawyers	241
Foreman of Riggers	1
Cabin Keeper	1
Riggers	103
Boatswains	21
Labourers	64
Foreman of Scavelmen	1
Scavelmen	118
Boys	30
Warders	36
Foreman of Labourers	1
Labourers	518
Storehouse Foremen	3
Labourers	37
Total Borne in the Dockyard	**3,749**
To which add:	
Officers and Men in Ropeyard	352
in Wood Mills	89
in Metal Mills	67
in Millwrights	70
Total Borne in the Yard	**4,327**

Whilst the dockyard largely undertook Navy and government work, there was also some repair work on merchant ships belonging to the great contemporary commercial ventures, particularly those of the East India Company, which used the yard between 1740 and 1805, as well as the Levant and Royal African Companies, and even – when Britain was not at war with their countries – the Dutch, Danish and Swedish East India Companies. In peacetime this arrangement was probably satisfactory, providing extra work and income for an underemployed yard, but in wartime, when the yard was fully stretched and materials were at a premium, any co-operation between the yard and the companies was given grudgingly. Between 1740 and 1760, some two dozen East India Company vessels received repairs, assistance and stores from Portsmouth's yard authorities. Such was the reliance placed on the yard by the Dutch East India Company that the United Provinces retained consular representation in Portsmouth[94] (using the house in Broad Street that is now the headquarters of Portsmouth Sailing Club).

Another source of work for the dockyard was the fitting out of naval ships built by local private yards: for example, between 1745 and 1814, fifty-three such ships were built at Buckler's Hard on the Beaulieu River, and each hull was towed to Portsmouth (by rowing boats, but making use of favourable tides), an 8-mile journey taking two or three days, for fitting out. Other private yards in the area that built for the Navy in the Georgian era included those at Bursledon, Southampton, Northam, Eling, Redbridge, Hythe, Warsash, Cowes, Fishbourne, Itchenor and Littlehampton.

The Late Georgian Period

Fully covered slips and docks were introduced at Portsmouth from 1814 onwards following the construction at Woolwich of a prototype wooden structure with elaborately braced roof trusses which oversailed their supporting pillars, designed by Robert Seppings, the newly appointed Surveyor of the Navy. The object was to reduce the occurrence of dry rot in ships, which had previously been built on slips open to all weathers. The roofs had glazed skylights and initially were covered in tarred paper, slates or tiles. After some years, zinc or copper sheets were adopted as standard.[95] The covers must also

have increased productivity, and were similarly used on those dry docks which did not need to take masted ships.

On Wednesday 14 September 1825, the 110-gun First Rate *Princess Charlotte* was due to be launched by Prince Leopold (after whose late consort the ship was named) at 12.30 p.m., but immediately before the event there was an unfortunate accident in the dockyard. *The Times* reported that 100,000 people were there for the launch, and they must have been crammed into every available space. Some were standing on the bridge above the gates at the entrance to the empty South East Dock, expecting to see the newly launched ship being towed into the basin for fitting out. There was an abnormally high tide and the water level in the Great Basin was 2ft higher than usual. The pressure of the water on the dock gates, which were said to be rotten, was so great that they gave way and a rush of water and debris was swept into the dock, taking with it most of the spectators on the bridge. The wave created hit the far end of the dock and rebounded along its length, washing some of the spectators into the basin. Immediately after this the launch took place, whilst people in the vicinity of the South East Dock, including many marines, rushed to try to rescue those who had been hurled into the water. The bodies of the injured and drowned were taken to the dockyard surgery, where seven were resuscitated whilst others were beyond help. The known death toll three days later was sixteen, including a midshipman from HMS *Victory*, a painter's apprentice (aged 15) and eight children (aged between 2 and 17), but it was feared that a further five or six bodies had yet to be recovered.[96]

The School of Naval Architecture, in South Terrace, was built in 1815–17, designed by Edward Holl, architect to the Navy Board. A long two-storey rectangular building with a pedimented centre, it was constructed of yellow stock brick in Flemish bond with limestone dressings (matching the Commissioner's House, which it faced), with a hipped slate roof with tall brick stacks, and originally provided classrooms and accommodation for up to twenty-five apprentice shipwrights. Established in 1811, and initially sharing accommodation in the Royal Naval Academy at Portsmouth, it was to take 'superior' apprentice shipwrights and train them in rigorous theoretical and scientific principles in the design of warships, something that the French navy had introduced as early as 1741. By 1860 its graduates held most of the senior technical posts in the Navy, and many outside. The school's authority was confirmed by the selection of a design for HMS *Warrior* by its graduate Isaac Watts, the Chief Constructor, in 1860. However, the school was closed during the reorganisations of 1832. The building was the forerunner of the dockyard apprentice schools and was subsequently put to a variety of uses.[97] Between South Terrace and the Commissioner's House is The Green, a garden created in the nineteenth century on ground previously occupied by saw pits, woodworking, timber storage and the house carpenters' and carvers' shops.[98]

By the end of the Georgian era, the Navy was on the cusp of revolutionary change as steam-powered vessels were being introduced. The Navy's first steamship, the tug and ferry *Comet*, was constructed at Deptford dockyard in 1822, and the first dockyard steam tugs at Portsmouth arrived in the 1820s. In 1832, as the Admiralty was taking its first hesitant steps towards making a marine engineering provision in the dockyards, under Goodrich's supervision, a boiler shop was opened in the Portsmouth yard for the repair and manufacture of both stationary and marine boilers, together with a millwrights' and engine-makers' shop for making steam engine parts.[99] In 1835, the first steam-driven warship to be built at Portsmouth was launched. She was the wooden hulled *Hermes*, a six-gun paddle sloop, designed by Sir William Symonds (Surveyor of the Navy, 1832–47) with 220hp engines by Maudslay & Field.[100] Portsmouth was the last of the home dockyards to start building steam-powered wooden warships, and would soon require more extensive facilities to meet the expanding needs of marine engineering and iron ship repair and construction.

Every day's work in a dockyard depends more or less upon the directions received by the morning mail from Whitehall. Hence it is that the correspondence at the Admiralty has become so voluminous, that important papers are forgotten and innumerable letters signed by persons who are totally ignorant of their contents.

Lord Brassey in a letter to *The Times*, 1872

The administrative structure of the Navy from 1689 onwards was headed by the Lords Commissioners of the Admiralty (the Board of Admiralty), a ministerial body whose composition varied over time but for most of the time comprised the First Lord and six subordinates. Prior to this, the Lord High Admiral had performed similar duties. Regarding dockyards, it mostly confined itself to making policy, delegating the running of them to the Navy Board (the Commissioners of the Navy), a body which actually predated the Board of Admiralty, having been established by Henry VIII in 1546, 'to oversee the administrative affairs of the naval service' (while policy direction, operational control and maritime jurisdiction remained in the hands of the Lord High Admiral). The Navy Board had the responsibility for building, equipping and maintaining the fleet in the dockyards, and for supplying the fleet with stores, on behalf of the Admiralty. Its offices (the Navy Office) were in Seething Lane, London, moving to larger offices in Somerset House in 1789. Headed by the Comptroller, who was selected from the senior captains in the Navy, it held regular meetings, usually to make policy, approve expenditure and justify its actions to the Admiralty. The members acted in a collective way, taking all decisions at their general meetings, which were held two or three times a week, or daily in wartime. The principal dockyard officers provided the Navy Board with information and, sometimes, recommendations, but could not make decisions themselves – because their role was to carry out the board's directives.[1]

The Navy Board's principal officers included the Surveyor, who was responsible for the design, construction and maintenance of ships, subject to the approval of the whole board. He was appointed from among the master shipwrights in the dockyards, and also had responsibility for the raw materials needed for building and repairing ships and for the design of buildings and the layout of the yard. Another, the Controller of Storekeepers' Accounts, oversaw the stores supplied to the dockyards, ensuring their quality and proper distribution, and receiving the regular accounts from the Storekeepers of each dockyard. The Clerk of the Acts (later known as the Secretary to the Navy Board) had voting powers on the board until the latter years of the Navy Board's existence, and was essentially a secretarial post, supervising the clerks of the Navy Office, with all instructions from the Navy Board passing through his office. The most well-known holder of this post was Samuel Pepys, who held it from 1660–73.

The resident (or yard) Commissioner of each dockyard was appointed by, and also a member of, the Navy Board. This office, which was invariably filled by a naval captain (whose career at sea was, often for health reasons, over) was established in 1690, principally as a channel of correspondence between the Navy Board and the principal officers of the dockyard, and for keeping the board informed on events in the yard. He rarely met with his colleagues in London and thus did not participate, except by correspondence, in the Navy Board's decisions. His role was ambiguous and his instructions unclear, at least until 1801, when some specific instructions were distributed, to the extent that the Commissioner at Portsmouth in 1749 had never seen his instructions. Instead, he had to base his conduct on the practice established by his predecessors, as evidenced by the records they kept. In 1801, the new instructions advised:

He (the Commissioner) is to make his constant residence in the dockyard in order to his always being in readiness to receive and see executed all orders from the Admiralty and commissioners of the navy relating to the service of the navy and not to absent himself from the post without permission first had to that purpose in writing from the Admiralty.[2]

Designed by Samuel Wyatt, Admiralty House was completed in 1787 as the Dockyard Commissioner's House. Facing The Green, it has spacious gardens to the rear and is now the residence of the Fleet Commander and Deputy Chief of Naval Staff.

officers. In the case of the latter, he could, though, reprimand them or recommend them for promotion. He received the daily letters from the Navy Board and was responsible for enforcing the board's regulations, orders and instructions, which were for this purpose sent to him to be delivered to the principal officers when they gathered together each morning to read the day's letters from the Navy Board and co-ordinate the activities of the various departments. The board addressed its letters to the dockyard officers collectively and issued both general warrants (standing orders) to provide policy and precedent, and ordinary warrants containing specific orders. However, the principal dockyard officers did correspond individually and directly to their opposite numbers on the Navy Board in London, though this was not the official protocol since all communications were supposed to pass through the Commissioner's office. Neither the principal yard officers nor the resident Commissioner could order work to be done in the dockyard without the full authorisation of the Navy Board, except in an emergency. However, in practice, for expediency's sake, they sometimes *did* authorise things like repairs to ships. Over time, a mass of standing orders accumulated, each issued, in Charles Middleton's words, 'to meet each present difficulty as it arose, perhaps without having the least reference to the general principle on which they had been at first established'.[3] Often the orders were contradictory, or reiterated or amended earlier orders, but no attempt was made to withdraw the earlier ones, and the resident Commissioner had to attempt to resolve confusions arising from this. From a figure of 200 standing orders in 1715, they multiplied exponentially so that by 1786 there were some 1,200 standing orders extant. Officially, all were supposed to be read out quarterly to the assembled yard officers, but unsurprisingly this was not complied with. It was not until 1806 that the standing orders were collated and codified.

There were frequent tensions between the Navy Board and the Admiralty, and the inefficiencies of the former were apparent to all. The First Lord of the Admiralty, the Earl of Sandwich, introduced general dockyard visitations by the Admiralty in 1749 as a means of monitoring the effectiveness of the Navy Board and improving the efficiency of the dockyards. The visitation of 1749 was only the second ever, the first having been conducted by the Navy Board at the behest of the Lord High Admiral, King James II, in 1686. There was another visitation in 1764, and in 1767 Sir Edward Hawke

Thus he had to ensure that the board's orders were followed, even though he did not have direct line authority over the yard officers. However, as the local representative of the Navy Board he usually had significant informal authority, though to some extent this depended on the Commissioner's ability to command respect. Given his background as a serving naval officer, he could not be expected to possess the same technical knowledge and experience as the yard's principal officers. Over time, the resident Commissioner had acquired the authority to appoint, discharge and discipline members of the industrial workforce of the dockyards, but not its

Far left: John Montagu, Earl of Sandwich, had three spells as First Lord of the Admiralty: 1748–51, 1763 and 1771–82. He concerned himself with the perceived inefficiencies of the dockyards, carrying out annual visitations which produced detailed findings for improvements. (Thomas Gainsborough)

Left: Georgian dockyard officers, a dockyard sentry and Samuel Pepys, head of the Navy Office.

Dockyard Sentry c1776 Regiment of Foot

Samuel Pepys of the Navy Board 1666

Dockyard Officer and son 1799

Dockyard Superintendent & wife 1830

ordered the Navy Board to visit the yards annually and report on what they found. Their reports were vague and ineffectual, so every year from 1771 to 1778 Sandwich (who had returned in 1771 to the First Lord post he had vacated in 1751) visited the yards accompanied by one or two of the junior lords of the Admiralty plus, from the Navy Board, the Comptroller, the Surveyor and the Clerk of the Acts. The visitations, each taking about two weeks, and the very detailed reports and orders arising from them, probably did lead to fairly marked improvements, but Sandwich's successors did not follow his example. Lord Howe visited the yards in 1784–85 and the Navy Board in 1792, but those were the only visitations until 1802. But the First Lord from 1812–27 and 1828–30, the second Lord Melville, firmly re-established annual visitations and his successors did not deviate from it.[4]

Sir James Graham, First Lord from 1830 until 1834, was frequently at odds with the Navy Board and its Comptroller, Sir Byam Martin, accusing it of financial mismanagement and of thwarting the Board of Admiralty, and retaliated by dismissing Martin and abolishing the Navy Board in 1832. Its functions were then taken over by the newly reconstituted Board of the Admiralty.[5] The dockyards came under the control of the Second Sea Lord, assisted by the Surveyor General, but authority rested with the Board of Admiralty alone. The Surveyor advised in his professional capacity, but did not have line authority over the dockyards.[6] The post of Admiral Superintendent replaced the resident dockyard Commissioner in July 1832, and acted as the local head of the dockyard and its link with the Admiralty. He was also the second-in-command of the port. But the authority of the post was strengthened because it included direct authority over all yard officers. Like the former Commissioners, the Admiral Superintendent was drawn from serving naval officers, normally a rear admiral, but unlike the Commissioner he did not lose his seniority. The daily meetings with the principal officers ceased in 1833, sacrificing useful communication and coordination. The Commissioner's House was refurnished as the residence for the Port Admiral and the house at the south end of Long Row was enlarged for the residence of the Admiral Superintendent.

The work of the yard was executed by salaried principal officers. In the eighteenth century they were the Master Shipwright (who was informally regarded as the senior officer, and was also known as the Builder), the Clerk of the Cheque, two Masters Attendant, the Clerk of the Survey, the Storekeeper and the Clerk of the Ropeyard. The Master Shipwright had two or three assistants, including the Master Caulker.[7] The Master Shipwright was a shipwright by trade (having himself usually been apprenticed to a master shipwright) and commanded the biggest part of the workforce in the yard, including shipwrights, caulkers, sawyers, house carpenters, labourers and scavelmen. He had overall responsibility for the building, repair, fitting and refitting of ships in the dockyard, including their launchings, dockings and undockings, and had oversight of the construction of ships in private yards. He was also responsible for the construction and repair of buildings within the yard. Gangs of shipwrights, or other artificers such as caulkers, were supervised directly by quartermen, who each reported to a foreman of their trade.[8]

The Master Attendant was a former master in the sea service and was responsible for activities afloat, including the care of the Ordinary (ships laid up in reserve), and the associated shifting of berths, storing, rigging and unrigging. The sailmakers and riggers of the dockyard came under his command, and he supervised a gang of shipwrights to maintain ships in the Ordinary. In many of these duties he had to work closely with the Master Shipwright. When ships were to be docked, the Master Attendant brought them to the dock entrance, where the Master Shipwright took over responsibility for them. The Master Attendant oversaw the launching of ships in private yards and brought them to the dockyard for fitting out. This could take him away from Portsmouth for some time and was one reason why two Master Attendants were needed. The Master Attendant also maintained all buoys and moorings and appointed pilots to navigate vessels in and out of the harbour. After the American War, the post of Superintending Master was created to take care of ships in Ordinary, thus taking some of the burden off the Master Attendant.

The Clerk of the Cheque kept the yard pay and muster books, checking the accounts of earnings and forwarding the pay books to the Navy Office, and held the yard contingency account, and he and his clerks had to muster the workmen at the start and finish of the day. He was informally regarded as second in seniority to the Master Shipwright. The Storekeeper was responsible for stores held for use in the dockyard, whilst the Clerk of the Survey dealt with those stores which were to be taken aboard ships that were being prepared for service or were in Ordinary. Jointly, these three Clerks were responsible for the reception of stores from contractors, with one of their staff attending the unloading and storing of these, accompanied by one from the office of either the Master Shipwright or Masters Attendant. An important additional function of the Clerk of the Survey was to check the stores issues and accounts of his fellow principal officers, but was hampered in this by his lack of seniority. The Clerk of the Ropehouse was not strictly speaking one of the principal officers, but did not report to any of them, so was similar in status. He was responsible for all activities in the Ropeyard, and for his own accounts and materials. In 1801, the posts of Timber Master and assistant were created to control the use and storage of timber, whilst a Master Measurer was appointed in 1810 to supervise the measurement work associated with piecework, the use of which had increased greatly during the Revolutionary and Napoleonic Wars.[9] There were many inferior officers who reported to the principal officers. As well as the Assistant Master Shipwrights and the Master Caulker, the Boatswain of the Yard, who supervised the scavelmen and labourers, the Master Mast Maker, Master Smith, Master Boat Builder, Master Joiner, Master House Carpenter and Master Bricklayer all reported to the Master Shipwright. An establishment list for 1813 also shows a Master Painter.[10] The Master Sailmaker and Master Rigger reported to the Masters Attendant; the Master Ropemaker reported to the Clerk of the Ropehouse; the other principal Clerks had senior clerks reporting to them. In 1688, a surgeon was appointed to the dockyard, providing medical services to the yard employees.

From 1652 until 1822, the hierarchy of supervision below the Master Shipwright and Assistant Master Shipwright was the

Foreman and below him the Quarterman. The Quarterman was a promoted shipwright, salaried from 1801, in charge of a gang of twenty shipwrights.[11] In 1822, the First Lord, Viscount Melville, adapted to the dockyards the system of supervision found in commercial yards, cutting a swathe through the supervisory structure. He dismissed the quartermen, who allegedly had little to do and spent much time loafing, thus breeding indifference and setting a poor example to the men. A leading man was appointed in each gang of ten shipwrights to receive instructions from the foremen of the yard (whose numbers were increased to nine at Portsmouth) and work alongside the shipwrights or caulkers in his gang: he did not have supervisory responsibilities and was not salaried.[12] Also dismissed by Melville were the foremen of trades, the Master Mastmaker and the Master Boatbuilder (with the

masters of other trades abolished in 1830). However, the foremen of the yard could not supervise the numbers of men involved and the salaried grade of inspector was introduced in 1833, with one inspector supervising a company (which consisted of two gangs) and inspecting their work. Two companies made a division, under a foreman of the yard. Like the foreman, the inspector was a promoted, salaried craftsman. The posts of Clerk of the Survey and Clerk of the Ropeyard were also abolished in the cuts of 1822,[13] followed on 30 March 1830 by the Clerk of the Cheque, their accounting duties being passed to the Storekeeper.[14] In 1856, the first Yard Accountant was appointed.[15] The Master Measurer's office, established in January 1811, and the Timber Master's office were abolished on 30 June 1827: on the following day, the offices of the Receiving Officer and Converter were created. On 31 March 1831, Simon Goodrich, who had been appointed Civil Engineer and Machinist in July 1815, was superannuated and his establishment broken up (though Goodrich continued to work in the capacity of consultant). On 1 July 1833, all the Measurers were superannuated and all the workmen in the yard were placed on day pay.[16]

The old order had changed further when wood and sail gave way to iron and steam. The steam Navy had a few tugs in the 1820s, and by 1837 consisted of twenty-seven paddlewheel vessels including two frigates under construction at Pembroke, and the Admiralty appointed a Chief Engineer in 1835. At Portsmouth, the Steam Factory opened in 1849 and was managed by a chief engineer, Andrew Murray, a marine engineer and former managing partner at William Fairbairn's Millwall yard. He was well versed in naval architecture and ship construction, and with management practices in private yards. He reported to the Admiralty's Engineer-in-Chief, Thomas Lloyd, within the Surveyor's department, and ran the Steam Factory with exemplary efficiency. The Committee of Revision, appointed in 1848 by the First Lord, the Earl of Auckland, to look into dockyard costs and management, had nothing but praise for Portsmouth's Steam Factory: 'There was not a man, out of 533 employed, whose work on any particular day … could not be at once ascertained and accounted for.' The Chief Engineer had equal status with the Master Shipwright (or Constructor as he became known), but the challenge to the latter's long-established supremacy gave rise to rivalry between the two men, and though good communication and coordination was vital between the two departments, it suffered as a result. The Master Shipwright would hurry to complete one ship, the Chief Engineer at the same time working flat out on the engines for another, neither knowing what the other was doing.[17]

The ability of the Admiral Superintendent to remedy such problems was questionable, because of his lack of technical training, and in 1867 the Controller of the Navy (a new title for the former Surveyor of the Navy), Rear Admiral Robert Spencer Robinson, observed that the Superintendent was responsible for nothing that was 'either done or left undone … so long as he duly transmits the memoranda from the Admiralty, the Lords, the Secretary, the Controller, the Storekeeper General, the Accountant General'. When a question was asked, he had to ask the relevant yard official to reply. Robinson proposed that each yard should have a civilian general manager to coordinate and control the constructive, engineering and store departments. The First Lord, H.C.E. Childers, introduced his own version of this in 1870 at Portsmouth and Chatham: the engineering and stores departments were subsumed into the constructive department, headed by the Master Shipwright, who acquired the additional title of engineer but lacked the relevant engineering competence. The Admiral Superintendent remained in post, to ensure that the Navy's priorities and wishes were fully responded to. A new post at the Admiralty, Director of Dockyards, was created in 1872 as a sort of travelling general manager, but the post holder became bogged down in administrative work at the Admiralty and did not visit the dockyards frequently. Childers' experimental local management structure in the dockyards was a failure and was reversed in 1875, but the Director of Dockyards post remained.[18]

In 1886, in a further attempt at solving the problem of coordination and control, naval constructors were appointed as civil assistants at the three main yards, supporting the Admirals Superintendent by giving advice on all the professional and technical aspects of dockyard work. They became quasi-general-managers and things probably did improve a bit, with the Admirals Superintendent becoming a bit more active, though the responsibilities of the chief constructor and the chief engineer in each dockyard were gradually usurped, and the post of civil assistant was eventually abolished in the Fisher reforms of 1905. Fisher's solution was to strengthen the roles of the chief constructor and chief engineer again, with new titles – Manager, Constructive Department and Manager, Engineering Department respectively – but naval supremacy survived in the form of the Admiral Superintendent, albeit with somewhat reduced powers.[19] The Director of Dockyards was relieved of his administrative routine in 1905 so that he could spend all of his time visiting the yards, accompanied by his engineer assistant, as a peripatetic general manager, but this was no substitute for resident general managers. A further step in decentralisation was taken in 1912 when the Director was empowered to issue orders on his own responsibility and to approve estimates for repairs, alterations and additions.[20] It was to be well into the twentieth century before a general manager was appointed to each dockyard, and the post of Admiral Superintendent was finally abolished on 15 September 1971.

Uncertainty over the most effective and economic way to supervise craftsmen continued in the late nineteenth century. In 1870, the grade of inspector was abolished as a cost-saving measure, but was reintroduced in 1884. In 1891, after an unsatisfactory attempt to incentivise craftsmen through a payment incentive scheme which allowed them to progress through four grades dependent on satisfactory commitment to work, the unsalaried grade of chargeman was experimentally introduced as a gang supervisor at Chatham, and it replaced the leading man in all the yards in 1898. Unlike inspectors and foremen, who were appointed by the Admiralty, chargemen were appointed by local management, were paid 1s per day more than the craftsmen and could be downgraded to craftsmen if they did not perform. The inspector would supervise several chargemen and be responsible for the workforce performance on any given refit or building project, freeing foremen to concentrate on the more technical side of the operation. Whereas at the beginning of the nineteenth century shipwrights had enjoyed considerable autonomy over their working lives, being able to decide the composition of their gangs at the annual reorganisation of the dockyard known as shoaling, the shoaling process was at the discretion of inspectors and leading men by 1890, and inspectors and chargemen after 1898; the craftsmen were more closely supervised and the upper management of the dockyards, the Assistant Constructors and Constructors, was more professionalised. The chargeman, inspector, foreman hierarchy became established as the basis of dockyard supervision into the Great War and beyond.[21]

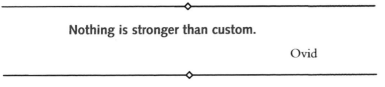

Nothing is stronger than custom.

Ovid

Theft

One of the greatest bones of contention in dockyards was the practice of employees taking home 'chips', supposedly off-cuts and other surplus wood that was of no use to the yard. What was regarded as a privilege by employees was frequently seen by the authorities as an abuse. The practice may well have dated back to Tudor times, and certainly by the late seventeenth century it had assumed unreasonable proportions at Portsmouth, and usable wood was often being cut up into short lengths so that it could be deemed to be chips and taken out of the yard.

The Lords of the Admiralty issued an order on 4 November 1674 directing that:

> no old wood ripped from any of his Majesty's ships, firing or chips of any kind shall from this day be enjoyed by any officer of the Yard by the name of a perquisite, or carried away by the workmen or people belonging unto this Yard, but to be laid in some convenient place for sale.

Fifteen years later, the abuse was still continuing, with the Commissioner, Sir Richard Beach, observing on 18 July 1689:

> It being grown a common practice every day by the shipwrights and others of this yard (almost all artificers) to carry out such burdens of chips that they can hardly stand under, not made lawfully, but split out on purpose in bundles to carry out of the Yard, to their Majesties' great loss of their time and waste of their stuff.

Beach subsequently ordered the Porter of the dockyard (who was in charge of the dockyard gate) to:

> use your utmost endeavours ... and ... take care not to suffer the workmen of this Yard to carry out firewood under the notion of chips ... [and] do not suffer them to carry out timber plank or board split, it having been too accustomary here that not only such as fall from the axe but what else they can get and split out although they do not a stroke of work the time they are in the Yard in rainy weather.

Further such warnings followed in 1695–97, though 'lawful chips' were to be allowed, these being those that had 'fallen from the axe' which could be carried loose under the arm. They should not be carried on shoulders or secured in bundles with cords or in bags, the penalty for which was discharge from the yard.

In September 1698, the Commissioner issued a notice allowing the poor of Kingston and Portsmouth 'to gather small useless chips' on Wednesday and Saturday mornings, but this again was abused and the Commissioner warned in the following month that the concession would be suspended if the rules were not followed. The beneficiaries of the concession to the poor were often not the truly needy, but were instead the 'sisters, cousins and aunts', or other relations, of dockyard employees who doubtless received chips that had been specially cut up for them. To control this, the Commissioner had in 1705 to obtain names of the really poor from the Mayor of Portsmouth and the Minister of Kingston Parish. But it seemed that however much the authorities tried to close loopholes, the problems with chips persisted. It had emerged in October 1700 that when the Porter removed unlawful chips from shipwrights, or all chips from labourers, sawyers and others who were not entitled to the privilege, these chips were later sent to the Porter's lodgings for fuel, rather than being returned to the yard. And in 1708, workmen had to be reminded that lawful chips had to be carried out under one arm without the assistance 'of the lappets or skirts of their coats'. In August 1783, further dress restrictions were issued by the Navy Board, which were apparently also directed at other types of theft: 'You are to suffer no person to pass out of the Dock gates with great coats, large trousers or any other dress that can conceal stores of any kind.' No person was allowed to wear a greatcoat and no labourer working in the storehouses could wear trousers.

A caricature of raucous Georgian life at Point, Old Portsmouth, on the wall of the Bridge Tavern, based on a painting by Thomas Rowlandson (1757–1827).

However, chips were a problem that would not go away. In April 1745, four labourers were discharged and had to forfeit one week's pay 'as an example to others' for cutting up chips between the piles of timber and in the teamers' stable. In July 1767, the Navy Board, following a visitation to the yard, observed that the privilege of chips was being abused and ordered that the rules be followed. In April 1770, the practice of allowing the poor to come into the yard to collect chips was stopped, and instead the chips were delivered to them. However, the beneficiaries were not happy with this new arrangement and 'were very troublesome in uproar and throwing stones etc over the Dock gates'. This disrupted the work of the yard, and the deliveries of chips were thereafter made to the North Gate, where there was more much room to carry out the distribution.

Things turned much nastier in September 1783 when large numbers of shipwrights and house carpenters refused to carry chips under their arms rather than on their shoulders. When this was reported to the Navy Board, it drew the response that it was the dockyard officers' fault for having previously 'granted indulgences', saying that any shipwright discharged for the malpractice 'will owe the severity of the example to your negligence in not executing with fidelity the orders you are entrusted with'. This, it was said, could lead to the dismissal of the officers 'if the King's service was in danger of suffering from the present conduct of shipwrights'. But before this communication could reach the dockyard there was an ugly incident after the shipwrights again refused to put their chips under their arms and some 200–300 people – men, women and children, including 'watermen and people of the place with many of the labourers' – began to heave brick bats and stones at Mr Maddox, Foreman of the Yard, and some quartermen. The quartermen seized two of the men responsible, but they were soon released by the mob before their identity could be ascertained. Some of the labourers threatened to murder Mr Maddox 'for taking away their chips', and in the evening when the men who had worked overtime left the yard, three or four men set upon him, and when the quartermen went to his assistance two of them were knocked down with large broomsticks. But some of the other quartermen seized a labourer, who was was imprisoned. When the instructions of the Navy Board were made known to those concerned, they seemed to have the effect of preventing further demonstrations of insubordination for a time. But in the following month a large number of people assembled with an effigy of Mr Thompson, one of the quartermen, who had recently disciplined two men for disobeying him. The rabble went 'in triumph with some hundreds of men, women and children' with the effigy to Thompson's house, where they burnt the effigy and broke the windows of his house, 'throwing fire into it and firing guns and pistols in the air'. The ringleaders of this escapade were subsequently discharged, which put an end to the disturbances.

In an order dated 29 June 1801, the Navy Board abolished the perk of chips and gave money allowances in lieu. Shipwrights and their apprentices in the last three years of their apprenticeship received 6d a day, shipwrights' apprentices in the first four years of their apprenticeship, caulkers, joiners, house carpenters and sawyers 4d a day, scavelmen and labourers 3d a day, and the apprentices of caulkers, joiners and house carpenters during their first four years 2d a day and 4d a day thereafter. This money was paid weekly, and removed the source of so much trouble, friction and fraud. No chips were allowed through the gates and the poor of the town no longer received this benefit.[1] The expression 'with a chip on his shoulder', for a resentful person, may well have entered common parlance from the dockyard practice described above.

Bundles of chips often disguised valuable items secreted within them, and indeed the theft of other items was common. In August

The Porter's Lodge was built in 1708 and was the residence of successive porters, who guarded the dockyard boundaries, prevented theft by dockyard workers and rang the muster bell. Workers could buy beer from the nearby Taphouse run by the Porter, 'to enable them to better carry on their labour and not to distemper them'.

1677, Henry Roy was suspected of stealing three coils of rope and sent to the gaol in Portsmouth pending the 'due course of laws'. In October 1710, shipwright Thomas Horton was found to have embezzled two old iron bolts worth 2s, and his punishment – as well as forfeiture of some of his wages – was to be discharged, followed immediately by being sent to sea on board the *Mary*. In January 1732, a boy, William Wood, was found taking iron out of the yard. His punishment was to be placed in public view at the Dock Gate with the bolts hanging over his shoulders 'on a Wednesday or Saturday, being the days appointed for the women to gather up the small chips' and to stand there at noon as the workmen went home for their 'dinner', in the afternoon when they returned, and in the evening when they left work. In 1767, two women, Elizabeth Swann and Susanna Mould, were prosecuted for breaking open a storehouse and stealing a quantity of oakum. They pleaded guilty and were ordered to be publicly whipped at the Dock Gate.[2] When women were involved in theft, it was usually when they came to bring breakfasts to the men of the yard, often leaving with chips in their baskets, within which other items were concealed.

Friends often attempted to shield others when passing through the dockyard gates, and sometimes large groups organised a rush past the officers. In December 1803, 'fifty or sixty ropemakers made a more than ordinary effort to push through the gates with the rest of the people.' About half of them were stopped and searched, the others making their escape. Amongst those searched were found quantities of hemp.[3]

Some idea of the scale of theft is given by the numbers discharged for embezzlement during the American War, from 1774–1783:[4] twenty-two in the period 8 February to 22 December 1774, eleven between 14 January and 31 July 1775, thirteen between 2 February and 14 November 1776, eleven between 2 January and 16 December 1777, five between 27 January and 26 June 1778, five between 23 April and 21 September 1779 and eight between 22 June and 6 December 1780. Labourers were the most common miscreants, though the crime was not confined to them, and the stores most susceptible to theft were coal, cordage, timber and, as its use increased, metal (usually in the form of iron and copper nails, but also copper sheeting). Petty pilferers were not prosecuted, because of the difficulties in securing convictions: instead they were dismissed and fined (an amount of three times

the value of the article was specified in an order from the Navy Board in October 1783).[5] The Winchester Assizes were feared, in contrast to the local town sessions in Portsmouth, where a weak magistrate and sympathetic jury would probably lead to acquittal.[6]

Crews of ships docked in the yard could also perpetrate theft in opportunistic ways as they passed in and out of the yard.[7] As well as petty pilferers there was planned theft, and particular offenders were the men from the transports bringing contractors' stores to the yards. On one occasion, in September 1782, over £50 worth of the king's cordage was found on a Sunderland brigantine collier at Portsmouth.[8] In some cases it could be difficult to establish who was responsible: when a consignment of shoes being delivered in June 1780 by the transport *Royal Oak* went missing, it was not known whether members of the ship's crew or the labourers tasked with unloading the goods were to blame. In January 1783, the Navy Board ordered that any loss by embezzlement committed on board naval transports was to be charged against the master's wages. Security in storehouses was often lax: in February 1779, twenty-one seamen's beds went missing, as did forty-eight sheets of copper when spread out to dry after painting. The Storekeeper reported, 'There is not a single lamp in the vast extent of the storehouses from the hulk to the south end of the new storehouse and only one watchman. This gives great opportunity to men from the contractors' ships nearby.'[9] The offenders who gave most trouble were the receivers of stolen goods, but they were sometimes tricky to pin down. One such was Edward Brine of Portsmouth. As a result of information, Commissioner Samuel Hood had his house searched in June 1779, found nothing, and reported to the Navy Board that the information was 'malicious'. However, in March 1780, following information from one of Brine's servants, Robert Martin, a quantity of copper was found at Brine's house. The Deputy Admiralty Solicitor at Portsmouth, Thomas Binstead, was convinced that Brine was 'by far the most capital receiver in this country, having sent melted down copper in the course of a few months to the London markets to the value of near £1,400, besides what he used himself'. The informer, Martin, was ostracised by the local community, and complained to Commissioner Hood that, 'in consequence of exposing my late master … the trades people and merchants of this and adjacent places treat me with a great degree of coolness and indifference; I may with propriety term it contempt.' Hood could not employ him in the yard for fear of disturbance, but recommended that he be paid 2s 6d a day subsistence money after being told he had been obliged to pawn his clothes. A year later he was in Winchester Gaol, having been found guilty of a counter charge by Brine of planting the goods in Brine's house. Not only was Brine acquitted at Winchester Assizes, but he won a civil case against his informer, with damages of £500. Brine was clearly a wily operator: according to Martin, he employed a coppersmith in pickling and cleaning copper sheets to remove identification marks, two blacksmiths in beating out the broad arrow marks from bolts and the yard founder for filing the mark from brass pieces.[10] Another receiver who escaped justice was Peter Henley of Gosport. Following information in December 1774 that he had a quantity of new rope, nails, bolts and old canvas in his warehouse, Commissioner Gambier sent the Master Attendant and Storekeeper with a search warrant, but the delay enabled Henley to dispose of the goods.[11]

In 1813, an anonymous dockyardman kept a diary, which included comments on incidents of theft. On 10 June, 'William Smith, commonly called Cherrypole, was discharged for being taken … with a piece of deal and a candle.' When his wife and a neighbour pleaded on his behalf, the officer in charge told them that he had been lucky not to get a more severe punishment. The very next day, another man, John Griffin, was arrested at the gate with 75lb of lead and copper, and was sentenced to death when he appeared at the assizes, although he was later reprieved and transported for seven years. In September there was another incident, 'One John Humphris, a labourer, was taken at the gate with eleven copper ends of bolts … some in two old shoes and some between two plates that his dinner came in.' The copper weighed 14½lb, and Humphris was arrested and taken to Portsmouth gaol. Soon afterwards two of his fellow workers were dismissed and imprisoned when pieces of copper were found under the floor of the room where they worked, presumably ready to be smuggled out when it was considered safe.[12]

As late as 1828, some vestiges of large-scale fraud remained: John Darby, a Portsea coppersmith and brazier who, with his wife, three sons and his employees, embezzled stores for resale in London, was sentenced to fourteen years' transportation.[13]

Security

In 1686, the Secretary to the Admiralty, Samuel Pepys, instructed the Dockyard Commissioners to take steps to limit embezzlement through inspection of men at the gates, control of the porters, nightly rounds of the dockyard and the supervision of watchmen. This led to the formation of a force of 'porters, rounders, warders and watchmen' to guard the dockyards. Porters identified and escorted visitors and accompanied them to the heads of departments, rounders patrolled the yard, warders were responsible for the keys and backed up the porters at the gates, and the part-time watchmen guarded buildings and areas by night. The watchmen were part-timers, being dockyard employees who remained in the yard, on a roster basis, every fourth night, after carrying out their normal work and received an extra shilling for the duty. The rounders seem to have been the senior branch of the force, as they kept an eye on the other three bodies and frequently reported their misdemeanours to the Commissioner.[14]

The watchmen were not trained for the work and many of them were quite unsuited by their character and integrity for any position

The dockyard wall (1711) in front of the Porter's Lodge (left) and the Police Superintendent's house.

of trust or responsibility. In 1678, an order from the Lords of the Admiralty required that ten watchmen be employed, 'taking their turns five of them half the night, manning five watch houses'; a watchman not on duty was to be in his bed. One watchman from the gate (where there were presumably additional watchmen) was to make rounds of the yard from the gate to 'the backside of the Anchor Smith's shop' to see that the watchmen were keeping good watch at their posts. The watchmen were to seize any suspicious person, and 'upon any occasion of resistance or danger they are to give the alarm to the whole yard'. They also had to ensure that no boats came near the wharves in the night, as this was a way of removing goods without passing through the gate. They were typically labourers, and, being drawn from the local community, were often reluctant to challenge or apprehend intruders. By 1689, some watchmen had become quite unreliable, being found drunk or asleep in their cabins or absent from their posts. For this they could be fined from their wages and warned that severe punishment would follow any repetition. They could also be fined for lapses in their duty: in May 1760, lead was stolen from the sloop *Peregrine* as she lay at the jetty head, and the Commissioner ordered that the watchmen be fined 5s each for lack of vigilance. In February 1719, the yard officers were ordered by the Commissioner to lock the Dock Gate at 10 p.m. and ensure that nobody went in or out except in extraordinary circumstances, which had to be closely observed. In 1739, it was recorded that the watchmen were each armed with a loaded and primed musket. In 1748, several of the watchmen were found to be hiring 'any wretch' to carry out their watch, paying them 9d from the 12d that they received as wages, so they got 3d a night for doing nothing. In August 1758, the Commissioner ordered that watchmen should not have a fixed station and would not know which station they were to watch at until they arrived for work each night, to prevent connivance and collusion.[15]

Theft could take many forms: in May 1714, furniture sold with the *Nonsuch* hoy was missing when she was delivered to the purchaser and an investigation was ordered. Ships in dock were vulnerable sites for theft because of the amount of materials being used and the number of workers employed on them. In the late eighteenth century, during working hours, the artificers were watched by warders, especially when they were employed with costly materials such as copper.[16]

A more satisfactory solution to the deficiencies of the watchmen had to be found. In October 1764, a marine guard was introduced for extra protection and the watchmen were discharged. But after six years the system was discontinued: the marines were drafted to ships fitting for foreign service and the watchmen reintroduced. In June 1783, with the end of the American War, the marines were brought back again, with a Captain's Guard complement of one captain, two lieutenants and 129 NCOs and privates. However, in December 1784 this was reduced to a Subaltern's Guard of thirty-six privates and NCOs, to serve as 'patrols only within the [dockyard] wall'. They were additional to 'warders, rounders and watchmen as formerly'. During the succeeding nineteen years, the marine guard was on several occasions withdrawn when men were required for service afloat.[17]

In April 1801, when food price inflation and discontent with wages were high, the Navy Board informed the Admiralty that 'depredations' on stores, especially at night, were 'more frequent than they used to be'. The Admiralty acted immediately to increase security at night, and the Port Admiral was ordered to provide guard boats to patrol the wharves throughout each night.[18] Early in 1803, the marine guard was replaced by a military guard from the land forces stationed in the Portsmouth Garrison. Their behaviour was not always exemplary. On the night of 18 September 1804, a captain of the Third Lancashire Militia, who was commanding the dockyard guard, tried to introduce women into the yard and was refused by the warders and watchmen. He sent a woman round by wherry to land at the King's Stairs, where he attempted to 'suborn and entice not only his own military sentinel, but also the two dock watchmen by offering them money to suffer her to land which they peremptorily refused'.[19]

By 1822, the watching arrangements showed signs of improvement, the Commissioner reporting to the Navy Office that the rounders were men of the very best quality selected from the shipwrights of the yard. However, it was a measure of their unreliability that the night watchmen were searched every morning by the gate warder in the presence of the 'Lieutenant of the civil watch', who was always a warrant officer of one of the ships in dock. The warders were 'the best we have ever had, who by their strict attention to this part of their duty have become very obnoxious to the people in general'.[20] There were between thirty-six and thirty-eight warders during the

1820s, recruited from the labourers and scavelmen of the yard, under the command of a civilian Yard Warden who was recruited from among the half-pay lieutenants of the Navy. At night there were eighty part-time watchmen, recruited from the daytime workforce and allowed to sleep for up to two-thirds of their watch so that they would be fit for work the next morning.[21] The complement of the dockyard military guard in 1824 was three commissioned officers, nine non-commissioned officers, two drummers and sixty-three privates.[22] The two systems (civil and military) gave sentinels who were unlikely to connive with each other.

In 1833, the civilian guard of the dockyard was reorganised and the Yard Warden was styled Director of Police. The new force was full-time and wore a distinctive uniform, topped off with a cloak and bowler hat. Meanwhile, the military guard stayed on.[23] Despite the relatively good conditions of service, the dockyard police forces were not an unqualified success, and eventually the Admiralty asked for an independent review of their efficiency. This task was allotted to Superintendent Mallelieu of the Metropolitan Police. He set about this in a thorough manner and his final report made it quite clear that, in general, the dockyard police forces were ill-trained and, in some instances, not devoting all their energy towards police duties. Recommendations were made to train the forces along the Metropolitan Police's lines, increase pay and introduce a system whereby men frequently transferred from one establishment to another to avoid undue familiarity with employees, many of whom were related to policemen of local origin. It was finally decided that the best answer was to disband the dockyard forces and introduce the Metropolitan Police into the dockyards. In fact, the Metropolitan Police already had experience of dockyard work, having taken over the responsibility for the two yards at Deptford and Woolwich, which were both already within the Metropolitan district, in 1841, but until 1858 they had been reluctant to accept any commitment outside the Metropolitan area.[24]

In 1860, the Metropolitan Police took over the duties at Portsmouth and proved to be very effective. Transferred from London, and living in barracks, they had no connections with the local community, and indeed were met with hostility and suspicion across the dockyard and the community at large. When a constable was assaulted by workers in Royal Clarence Yard, the Gosport magistrates refused to convict. In the first year of their service, the

The Police Superintendent's house faces the Porter's Garden. It was built in 1908, at a time when the Metropolitan Police were responsible for policing the dockyard.

Metropolitan Police detected 645 persons pilfering, compared with a mere sixty-five under the old Dockyard Police. They also prosecuted forty-two of Portsmouth's marine stores' dealers and claimed that another eighty-three shopkeepers had discontinued business. They were responsible for policing dockyard and naval personnel within the dockyard and a 15-mile radius around it (with the proviso that these powers were only to be exercised in respect of property of the Crown or persons subject to naval or military discipline). The Admiralty Powers Act followed in 1865, which gave considerable assistance to the police in that it made the Admiral Superintendent of each yard a magistrate – thus giving him the authority to hear cases brought before him by the Metropolitan Police. Cells were built adjacent to the main police offices, and offenders were kept in them overnight after being charged and brought before the Admiral the next day. The Admiral could, if he found them guilty, impose a fine or term of imprisonment. As magistrates, the various admirals were also able to issue search warrants, which again made life much easier for the police. In 1908, a Police Superintendent's house, 19 College Road, was built in red brick on the east side of the Porter's Lodge in part of the Porter's Lodge garden. Like the Porter's Lodge, it is now used as offices for Portsmouth Naval Base Property Trust.[25]

From 1923 onwards, the Metropolitan Police in the dockyards were gradually replaced by the Royal Marine Police Force to relieve the pressure on Metropolitan Police manning. All the members were originally serving and retired members of the Royal Corps of Royal Marines and were subject to military law under the provisions of the Army Act. Pensioners from the Royal Navy were subsequently and reluctantly also accepted. A Chief Constable was appointed in 1932, but the replacement of the Metropolitan Police was not completed until 1934, due in part to the Admiralty's desire to proceed cautiously. In the case of Portsmouth, the change came about in August 1933. During the Second World War, serving members of the armed forces were retained when they reached retirement age so that the Royal Marine Police Force was deprived of recruits. The Admiralty Civil Police was formed as a stopgap, its recruits being exempt from joining the armed forces. The situation was regularised in 1949 with the formation of the Admiralty Constabulary, which took over all police duties.[26] The Ministry of Defence Police absorbed the Admiralty Constabulary in 1971, along with the Army Department Constabulary and the Air Force Department Constabulary.

Rioting and Rebellion

On 31 December 1742, there was trouble with the shipwrights and their apprentices who objected to the severity of the fines imposed for absenting themselves without leave. The Master Shipwright stopped a day's pay whenever they were absent from their workplace without leave, 'a practice they are too much addicted to'. They assembled in a body at the dock gate 'in a riotous manner' and would not let other workmen enter the yard. The Commissioner exhorted them to return to their duty but they would not, and by noon 'they had drunk freely'. The Mayor of Portsmouth was summoned to read the Riot Act, but it was a long time before they dispersed. The following day they assembled again and behaved in a similar manner. Marines were sent from ships in the yard to prevent any accidents to the magazines, and one person was taken into custody. The riotous behaviour continued that day, and in the evening they barricaded the gates with two boats (one of them being nailed to the gates) and a wagon load of hay, to prevent men leaving the yard to go home, threatening to attack them if they attempted to do so. These workmen had to be kept in the yard all night, two bags of bread and some cheese being obtained for them from one of the ships in dock. The following morning the

A shipwright boring holes with an auger. (From R. Dodd, *Days at the Factories*, 1843)

rioters forced their way into the yard and rescued the man who had been taken into custody. In the meantime, the shipwrights sent a petition to the Admiralty for the redress of their grievances. Further rioting took place on 24 January 1743, with violent behaviour, and a military force was summoned to suppress the riot. Five men were arrested for making speeches tending to foment trouble. As a result of the rioting, the Admiralty discharged thirty-four dockyard men 'who were very active in the riot'. However, as no further rioting was recorded it is probable that some concessions were made to the requests of the men.[27]

In October 1745, the officers of the yard were ordered by the Commissioner to take every precaution against insurrection by Jacobite supporters of Bonnie Prince Charlie, the Young Pretender, who was supported by the French in his claims for the throne. The country was at war with France and Spain, and a French invasion was feared. The officers were told to preserve the magazines from fire or any other incident and to prevent the embezzlement of stores.[28]

Beer and Tobacco

In the late seventeenth century, if not earlier, workmen were able to buy beer during working hours from the Taphouse run by the Porter and situated next to his living quarters, from which he probably made a healthy profit, but the privilege was open to frequent abuse. In May 1689, the Commissioner instructed the Porter to stop selling strong liquors, which he said was 'debauching workmen and others to the ruin of their families' and causing 'other material disservice'. Three months later, he wrote to the Clerk of the Cheque asking him to ensure that the practice of selling strong beer was stopped, 'it being a common practice for the men to be drunk with it'. He explained that the Taphouse was appointed to 'give a pint of drink to a man at a time when dry and that but middling and not strong beer'. The Clerk was to taste the beer to see if it was strong or middling, and search the rooms and cellar to find men drinking and tippling when they should not be, in which case they would suffer a deduction in their pay. In November 1689, the Porter was instructed to stop selling brandy and other liquors, it having only just come to the Commissioner's attention that this was

happening, with 'the workmen neglecting their time and wasting their own substance to the ruin of their bodies and families'. An exception was bottles of wine, which could be sold to 'gentlemen and strangers that come into the yard'. However, the malpractices crept back in, for we find the Navy Board writing to Mr Jones, the Porter, in April 1697, threatening him with suspension if he did not cease the sale of wine and strong beer. In December 1701, the Commissioner ruled that workmen should no longer be allowed to enter the Taphouse and should instead be served at the door, and they should not 'lurk or harbour themselves' there. Again this was soon being ignored: in August 1705, some quartermen and their servants were discharged for drinking in the Taphouse and others were fined one week's pay. The Taphouse continued to operate throughout the eighteenth century.[29]

One group of workers who were entitled to receive beer were the smiths, when making or repairing anchors, in view of the arduous nature of their work. In March 1728, the Commissioner wrote to the Clerk of the Cheque and the Master Smith, instructing that each such man should receive one quart of strong beer a day, to be increased to three pints a day when making anchors of 40cwt or more. This beer was purchased from the King and Queen public house, on the Common Hard, Portsea.[30]

The last Porter of the Gate died in 1801 and his successor, appointed as Warden, was a lieutenant. A person known as a Tapster was appointed to take charge of the Taphouse. In December 1832, the Admiral Superintendent reported to the Commander-in-Chief that a shipwright had been found in an intoxicated state in the Taphouse and had assaulted two Warders when they tried to get him to leave; the shipwright was discharged. Soon after this the Taphouse was ordered to be closed and the privilege abolished.[31]

On 15 March 1663, an order was made strictly prohibiting the smoking of tobacco in the yards and on ships in dock. This was over seventy years after smoking was imported to England from Virginia by Sir Walter Raleigh in 1586. The order was reinforced by an order from the Commissioner at Portsmouth in August 1694, saying:

tobacco not only creates a neglect of duty but also exposes (the King's) magazine here to the fatal consequences of fire, yet it is very much used and several skulk in holes and corners to prevent their being discovered in that forbidden and evil practice.

Offenders would be fined three days' pay for the first offence, double for the next. Foremen found guilty of not enforcing the order would be fined one week's pay. The prohibition of smoking in the dockyard remained in force for many years after this.[32]

Hogs and other Livestock

A picture emerges in reports during the eighteenth century of hogs and other livestock running free in the dockyard, whilst workmen habitually brought their dogs into work, sometimes with unfortunate consequences. In the period 1703–59, the standing order prohibiting livestock from running loose in the yard had to be issued six times by successive commissioners. It seems that each time a new commissioner was appointed he had to try to enforce the order, his predecessor having tried to no avail. In December 1703, the Commissioner's order to the Yard Officers noted that the hogs were 'said to belong to someone or other of the [dockyard] officers' and they were told to keep them in 'such places as you have proper for them'. In July 1714, the Commissioner complained of hogs running loose in the dockyard and getting into the dwelling houses and gardens of the officers, as well as the 'interruption and hindrance to Her Majesty's works by their nastiness and running amongst the workmen'. In July 1729, the Commissioner complained that the 'routing and turning up the earth render it not only dissightly and inconvenient', but 'ill nature may construe it a farm rather than a dockyard'. This description may have been accurate since not only hogs, but goats, sheep and rabbits were mentioned. As late as December 1814, the Commissioner had to write to the Yard Officers:

> You are hereby required and directed from and after the 15th of next month, not to permit any fowls, rabbits, pigeons or other livestock whatever to be kept in any of the cabins, stables or other parts of this Dock (except at the houses and stables of the resident officers).[33]

In December 1749, the Commissioner had cause to write to the Clerk of the Cheque following two incidents in which the dogs owned respectively by one of the ropemakers and one of the smiths had bitten other workmen, one dog being 'mad' and the other 'suspected of being mad'. All workers, and the chip women, were to be told that they were prohibited from bringing dogs into the yard, a practice that was then commonplace. Any person disobeying the order would be discharged, and the Porter of the Yard Gate was instructed to drive out or kill any dog found in the yard.[34]

False Identity: the Female Shipwright

Mary Lacy, born in 1740 to a working-class family in Wickham, Kent, left her skirt and bonnet behind and slipped into breeches and waistcoat at the age of 19, called herself William Chandler (Chandler being her mother's maiden name), and enlisted aboard the ninety-gun ship *Sandwich* at Chatham in 1763. She became the servant to the ship's carpenter, Baker, who was a kindly man but a violent drunkard. Mary underwent many hardships onboard ship, including a serious attack of rheumatic fever. Worst of all, Baker had fallen into drink and stopped paying her, if he had ever paid her at all. In the autumn of 1760, her rheumatic complaint was so bad that she ended up in hospital at Portsmouth, and was then assigned to the *Royal Sovereign*. She was released from the Navy in 1763 with the end of the Seven Years' War. Mary remembered that, 'On this occasion, my joy was so great that I ran up and down scarcely knowing how to contain myself.' But she did not go home, having decided that she wanted to become a shipwright's apprentice. In March 1763, she was signed on as apprentice to Alexander McLean, the acting carpenter of the *Royal William*, which had been decommissioned at Portsmouth Dockyard as a living ship. She served her apprenticeship in the dockyard and received her shipwright's certificate in 1770. 'William Chandler' became friends with a number of ladies, which was lucky for Mary because it probably saved her from being found out. Mary had written to her parents all about her adventures. A family friend (who maybe wasn't such a good friend after all) was let in on her secret and began spreading rumours around the dockyard, which prompted two of the men to approach Mary to speak about this. Despite being thrown into 'a most terrible fright', she managed to laugh this off, but two of the shipwrights that she worked for took her aside and demanded the truth. She broke down and admitted she was in fact a woman. Out of respect for Mary, these men swore to keep her secret and convinced everyone else in the dockyard that Mary was indeed a man, citing her interest in so many ladies as evidence.

In 1771, she was again struck down by rheumatism. By this time both her parents had died and she had no one to turn to for help until a family friend, a Mr Richardson in Kensington, who was apparently aware of her situation, invited her to stay with he and his wife, where he helped her apply for a Navy pension. He applied in 1771 under her real name, and the Admiralty minutes dated 28 January 1772 noted:

> A Petition was read from Mary Lacy setting forth that in the Year 1759 she disguised herself in Men's Cloaths and enter'd on board His Maj. Fleet, where having served til the end of the War, she bound herself apprentice to the Carpenter of the *Royal William* and having served Seven Years, then enter'd as a Shipwright in Portsmouth Yard where she had continued ever since; but that finding her health and constitution impaired by so laborious an employment, she is obliged to give it up for the future, and therefore, praying some Allowance

for her Support during the remainder of her life: Resolved, in consideration of the particular Circumstances attending this Woman's case, the truth of which has been attested by the Commissioner of the Yard at Portsmouth, that she be allowed a Pension equal to that granted to Superannuated Shipwrights.

Mary was granted a pension of £20 per annum as a superannuated shipwright without delay, despite giving her real name in the papers. She collected her money in Deptford and there met a shipwright named Josias Slade, and married him in October 1772. In the following year she had the first of her six children.[35]

Misconduct

It is perhaps not surprising that there were many reports of misconduct in the yard, given the size of the workforce, and instances such as absence, idleness and insolence provide illustrations of the problems experienced. In January 1694, workmen on the new Third Rate and Sixth Rate were visited at their stations by Captain St Lo, the acting Commissioner in the absence of Captain Benjamin Tyrell. He rebuked them for idleness, at which they 'hollered at me in derision, and impudently repeated the same from one to the other'. The men, who may have resented the criticism and reprimand of a *locum tenens*, were fined one week's pay, whilst the foremen of the Third Rate and the quartermen of both ships were fined fourteen days' pay.[36]

Although it was customary to write off as 'run' men who absented themselves from their work without leave, it seems they could generally get reinstated if, within a reasonable time, they expressed penitence and good intentions for the future. One such man was Isaac Palmer, shipwright, who, the Commissioner said:

> has been so deluded and drawn away by idle company which has had so unhappy an influence on him, but being truly sensible of, and sorrowful for, his fault, humbly prays he may be restored to his business … he promises to make amends for his past neglect by his future diligence and that he will not be guilty of the like for the time to come.

A model of the *Royal William*, on which Mary Lacy signed on as an apprentice in 1763 under the guise of a man.

The Commissioner instructed in June 1712 that Palmer be re-entered on the books of the yard.[37]

In August 1730, the Commissioner had to warn workmen of the yard of the consequences of 'leaving the works of this yard to follow their own inclinations of assisting merchants, builders and harvesting as well as other sorts of labour, for the prospect of some little advancement to their wages'. It seems that some workmen had come to regard their employment in the yard as a stand-by job which they could fall back upon when there was not any more pleasant or remunerative work available elsewhere.[38] The yard officers themselves were not always blameless. In October 1704, the Commissioner complained that the yard officers 'do daily employ a considerable part of the labour as well as other workmen, in their domestic business, going on errands and to the market', and instructed them to redress the situation or they would 'answer to the contrary'.[39]

One repeat offender was labourer Robert Macklaine, described as 'an idle negligent fellow', who frequently left the yard after the muster call and was recommended for disciplinary action by the Master Builder in January 1711. He proposed that Macklaine should have one year's wages deducted, the inference being that payment of his wages was in arrears by at least that length of time. Such a severe punishment may seem, in the circumstances, unreasonable. In order to subsist for at least a year without wages, he would have had to get credit how and where he could from shopkeepers and his landlord, probably by selling his wages at a discount for payment to the creditor at a future date. However, the Commissioner was not sympathetic, and, ordering an even more harsh punishment, instructed that Macklaine be discharged and should forfeit all the wages he was due.[40] Until 1805, dockyard wages were not paid promptly or regularly, and workers often had to suffer hardship and mortgage their wages at exorbitant interest rates. Men ran up high levels of debt, which sometimes caused them to 'run', that is disappear, in fear of being arrested for debt. In October 1697, it has been estimated, wages due to artificers represented about one and a quarter years' wages. The plight of workers can be gauged from a letter from the Commissioner to the Navy Board:

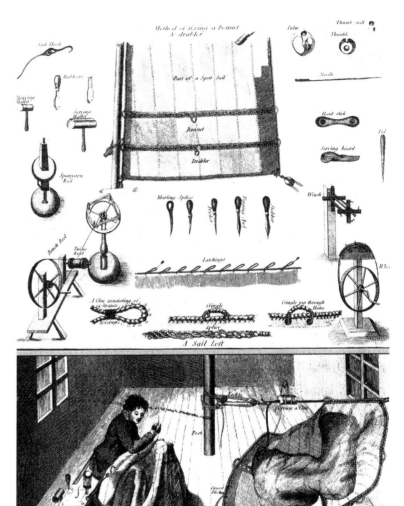

A sailmaker and his tools. (From David Steel, *Elements of Mastmaking, Sailmaking, Rigging, etc,* 1794)

Your late kind promise of reminding the Treasurer of the Navy for half a year's wages for the Yard put new life into the workmen; and a spirit of vigour and satisfaction seemed to be revived in them; but their wants daily increasing, and their credit declining more and more, they could not forbear representing their unhappy circumstances to me by petition, which has been also seconded by another from their creditors themselves, setting forth the hardships they are reduced to for want of money to carry on their trades, insomuch that many of them are forced to shut up their shops.[41]

In October 1729, the Navy Board listed a catalogue of misconduct and fraud which it said was commonplace in the Sail Loft. The sailmakers would complete what was considered, by custom, a reasonable amount of work (known as a 'stent') and then carry out work for themselves, making breeches which they sold to shipwrights and seamen. They often went out of the yard with these breeches on (presumably over their normal attire), returning without them on. The servants (apprentices) to the sailmakers were:

generally at play and idle away the King's time and do little but mischief, and make it a practice to carry out old ropes, tabling, and anything they can pick up in the Sail Loft and bind them up in bundles with twine which they carry out of the Yard under their coats, and that their masters take no care of them until they are near out of their apprenticeships, and then learn their trade.

Sails were seldom repaired, but were instead taken away at night for private use, having been allocated (together with bolt ropes) in free raffles, in lots, to the sailmakers. All of these irregularities were apparently condoned by the Master Sailmaker.[42]

It was customary for heads of dockyard departments to make personal gain by accepting payments for the entry of clerks. This privilege was abused by the Clerk of the Cheque, Mr Snell, in 1781, as highlighted after his dismissal of Thomas Morgan. Snell was paid the considerable sum of £350 in 1775 when Morgan was 'placed by the bounty of a deceased and much to be lamented friend, as the established sixth clerk' in the Clerk of the Cheque's office, and Morgan 'concluded it was a provision for me during my life'. About five years later, Morgan was promoted to fifth clerk, but in late 1780 Mr Snell dismissed him. Morgan appealed to the Navy Board,

requesting that Mr Snell be directed to return part of the money, and the Board instructed Snell to redress him, adding, 'never take upon yourself in future to disobey our order on this or any other head without acquainting us with the reasons that prevent it'.[43]

Disciplinary Code

By 1912, the basic disciplinary code was as revealed in a memo[44] from the Admiral Superintendent to the Manager of the Constructive Department, which listed the punishments for breaches of the code:

Offence	Punishment
Having matches	2 days suspension
Smoking and having matches	2 to 6 days suspension
Idling	1 to 2 days suspension
Betting	One month suspension or discharge
Losing time	One month suspension or discharge
Incorrect statement of work performed	One month suspension or discharge
Unlawful possession of government property	Discharge
Absent from work	2 days suspension

Punishments for the first three offences were to be increased in severity, at the discretion of management, with repetition. Rear Admiral Herbert, Admiral Superintendent in 1885, was convinced that dockyardmen were exceptionally idle, and that their supervisors tolerated this. It was alleged in the national and local press that dockyardmen engaged in organised avoidance, with members of the gang being detailed to keep look-out for supervisors, a duty referred to as 'keeping crow', whilst the rest of the gang idled. Another Admiral Superintendent, R.F.A. Henderson, felt that the main cause of idleness was 'a general disposition on the part of the men to do only what they considered sufficient for the wages paid', reflecting their resentment of the fact that their wages were lower than those of the equivalent worker in a private yard. But the dockyardmen also resented the service-style discipline, which they

Dockyard workers leaving the Main Gate, c.1910. This gate was widened in 1943, when the wrought iron arch and lantern between the piers were removed. It is now known as the Victory Gate.

said the naval officers at the top of the dockyard and Admiralty hierarchy expected of a civilian workforce. This was particularly true in times of full employment, when the compensations of dockyard job security were less apparent.[45]

Despite the efforts and claims of the Metropolitan Police, theft was still a problem in the late Victorian and Edwardian eras, and attitudes prevailed that petty stealing of tools or materials was somehow a dockyardman's right. In 1911, a riveter named Edmund Bartlett was arrested for taking an oilskin coat of Navy issue out of the dockyard, having taken it from HMS *Bellerophon*, the ship he was working on. His defence, supported by his chargeman, was that it was common practice for dockyardmen to take from ships clothing or gear which had been discarded by sailors. He had no intention of stealing from the Admiralty, or knowledge that he was committing an offence. The Portsmouth Magistrates accepted this defence, and Bartlett was acquitted. The Admiralty's response was to issue a general order prohibiting all taking of articles from ships, to be posted throughout the dockyards. Convictions for petty theft did occur: for example, on one day in 1913 three dockyard labourers were convicted of taking metal, engine fittings, copper pipe and fishing line, in two separate cases.[46] Recording of work was another problem area, with skilled labourers employed on piece-work. Recorders of Work had been introduced on 1887, one for every four gangs, but they and the gang supervisors would sometimes present false figures, perhaps to give the impression of greater efficiency than was the case, or as a means of disciplining the men under them, or in tacit agreement with the skilled labourers to bring wages in line with their conception of a fair return for the work performed. A trade union official, T. Sparshatt, was charged with recording sixteen rivets which he had not drilled. He did not deny this, saying that he had been engaged in awkward but necessary work in which it was impossible to make the piecework scheme pay; consequently he had charged what he regarded as a fair volume of work. One former skilled labourer alleged that misrecording was commonplace, and used to reward favourites or 'balance the books'.[47]

Day, Task and Job Work

In the eighteenth century and early nineteenth century, dockyard labour was much less productive than in private yards, according to the dockyard historian Haas, and the productivity varied widely, depending on which master shipwright was responsible for the work: some master shipwrights simply did not concern themselves very much with the cost of the work entrusted to them, and did not manage their workforces effectively. Until 1775, shipwrights were on day wages, whilst private yards were more likely to use task wages, i.e. payment by results, and day wages did not promote productivity. Another reason was said to be that the dockyards had a large number of inefficient shipwrights, who might be old, infirm or lazy, who would find no place in the private yards. In 1775, when task work was introduced in the dockyards, 41 per cent of shipwrights were excluded from it, on the grounds that they were too old or lazy. The strength of the gangs was equalised by 'shoaling': the yard's shipwrights were assembled and the quartermen took turns picking a man until all the gangs were up to strength. In contrast, in private yards the men of each gang could reject any prospective workmate, or expel a workmate, who was not as strong or hard-working as themselves, so inefficient shipwrights were apparently virtually unknown in private yards.[1]

When, in 1752, the Admiralty complained about how little work the shipwrights did, the Navy Board replied that they did as much as could be expected considering how much they were paid, 2s 1d per day, a rate that had not changed since 1690, and was about as much as was paid at the West Country ports which provided Portsmouth and Plymouth with many shipwrights. However, shipwrights employed by private yards in the Thames and Medway area received about 3s 6d on day-rate, which was more than the Admiralty paid a foreman. On task work, the private yards' shipwrights could earn up to 7s a day in wartime and about 4s 6d in peace time. The shortfall in dockyard shipwright earnings was often made up by overtime – which shipwrights constantly petitioned for – an approach that did nothing to improve productivity, since ten hours of work might be stretched out to thirteen or fifteen. Overtime rates were double the time rate per hour, and these earnings were seen, even by the Navy Board, as an inducement for good shipwrights to enter the dockyards. The standard working day, excluding overtime, was largely determined by the number of daylight hours. From March to September, dockyardmen worked from 6 a.m. to 6 p.m., with time for breakfast and two hours for 'dinner' at midday (allowing them to go home for it). In December and January, they worked from 7 a.m. to dusk, with one hour for dinner, taken in the yard. The men were mustered two or three times a day – morning, after dinner (if they were allowed to leave the yard) and in the evening – a process which could take up to one and a half hours. Another way of augmenting wages and salaries was through the supervision of apprentices, which attracted a premium and the receipt of a large part of the apprentice's wages. Quartermen, and those above them, were entitled to one or more apprentices, leaving less than 20 per cent of the apprentices to be articled to craftsmen.[2]

Superannuation was introduced by the First Lord, the Earl of Egmont, in 1764 as a way of getting rid of incapacitated workers who were often kept on by the yards for humanitarian reasons. It was limited at first to just 2 per cent annually of quartermen, shipwrights and caulkers. Workers who had thirty years of unbroken service qualified for two-thirds pay, whilst those prematurely retired because of permanent disability would receive a pension pro-rata to their length of service. Half of the men superannuated in the first group of retirees were in their 70s and 80s. At Portsmouth, in 1752, there were at least three men still working in their eighties, and at least five in their seventies. One Plymouth shipwright, aged 85, had sixty-four years' service and was described as 'very Lame

A view of the dockyard, looking across the Middle Storehouse (No. 10) towards the north-east, in 1911.

in both legs, Eye Sight impaired, quite enfeebled'. In 1771, Lord Sandwich increased the proportion superannuated each year to 2.5 per cent and extended superannuation to all dockyard occupations. The problem of aged and infirm men was alleviated, but was by no means eliminated because of the limit on numbers and the fact that the scheme was poorly administered.[3] Also, there was no mandatory retirement age until it was fixed in 1858 at the age of 70, and many elderly men were kept on just to increase their pensions. The age limit was lowered in 1864 to 60 for established men, although hired men could serve until the age of 65.[4] Private yards did not pay pensions, and the work offered was often highly cyclical and weather dependent, so wages could drop significantly and men could find themselves destitute in lean times when work was scarce, so the relative job security offered by the dockyards was an attraction. The dockyards did usually lay some men off when wars ended, but sometimes this only extended to a few indifferent hands. Dockyardmen injured on the job received free medical attention and were paid in full for the first six weeks they were off work, and at a reduced rate thereafter until they recovered, privileges unheard of in the private yards.[5]

The introduction of task work was seen as a solution to the problem of low wages because it increased wages substantially whilst leaving the underlying time rate unchanged, and at the same time eliminated overtime. It was introduced for shipwrights in 1775 by Lord Sandwich, who had already applied it to joiners, bricklayers and house carpenters in 1772–74. Certain kinds of unskilled labourers' work had been performed by task since 1758, whilst sawyers and rope makers had long been paid on task. For shipwrights, the Surveyor drew up a scheme which divided ships into twenty-five stages of construction, called articles of task, from keel laying to launch, and established a price for each class of ship and for each of the articles, to be used for shipbuilding and major repairs. Initiating the scheme at Portsmouth, the Navy Board wrote to the yard officers on 23 March 1775, 'As *Lion* is in frame and her time for seasoning is finished, complete her by task.'[6] Men who were either old or injured were excluded from the task gangs, and the gangs worked on task work in rotation, whilst time (day) rates were paid for other types of work such as minor repairs. Forty-two gangs, each of about twenty men and apprentices, were formed, of which twenty-one were task gangs; there were ten day gangs of shipwrights, and one of caulkers; the nine gangs employed in the mast house, boat house and capstan house were apparently also kept on day rate.

The yard officers were told that the task gangs should not work overtime. After two months' experience of the scheme, the shipwrights chose the occasion of Sandwich's visitation to Portsmouth to strike, complaining that task work was too hard for the prices paid. They also resented the loss of overtime and feared that their number would be reduced if the scheme was successful. The strike spread to Plymouth, Chatham and Woolwich, and the disruption lasted two months, with many shipwrights discharged because of absence from work, until starvation and the indictment of the ringleaders at Woolwich for conspiracy to procure an increase in wages, a Common Law offence, caused them to return to work, though the ringleaders (at Portsmouth and elsewhere) were refused re-entry. The shipwrights had been earning 4s–4s 5d a day on task, compared with an effective time rate of 3s 4d. Sandwich soon agreed that the prices were too low and the scheme was revised, increasing average earnings to 5s 3d per day, and task work became voluntary. At Portsmouth, most did not volunteer and the problem of the productivity of day gangs returned.

With the pressure of work during the American War of Independence, master shipwrights were given the authority to improvise piecework for minor repairs, for 'jobs' each of which was some part of an article of task work, at 'prices' which they themselves set. This inevitably led to wages drift and the system was discontinued when the war ended. The time rate was reduced to 2s 8½d in summer and 2s 1d in winter, at which the shipwrights again protested and asked for job work (and even the reviled task work) to be reinstated. The Navy Board agreed to restore piecework (job), but set a ceiling on earnings of 3s 4d (summer) or 2s 8½d (winter) to limit the effect of excessive prices.[7] At Portsmouth, shortages of shipwrights persisted throughout most of the American War: in March 1780, the Navy Board ordered forty shipwrights to be sent from Plymouth and twenty-one from Chatham, as well as thirty-one caulkers from Deptford and Chatham for the summer months.[8] When hostilities ended in 1783, the Portsmouth shipwrights finally accepted task work because overtime was cut[9]

At the start of the Revolutionary War in 1793, basic day rates were still those dating from 1690 and there were considerable shortages of men, mainly because the private yards paid so much more. However, job and task work in the dockyards meant that considerably more could be earned, up to a ceiling of two or three times the day rate. This was necessary because between 1770–95 the price of bread doubled, whilst other consumer goods increased in price by about 50 per cent; earnings were out of kilter with the cost of living, a situation mirrored in the Navy, leading to the seamen's mutinies of 1797. In 1794, because the trades working 'by the piece' were so often given the ceiling of 'two for one', two days' pay became standardised as the usual amount of pay (though this was not formalised as a new day rate until 1812), overtime also being added. Harvest failures in 1799 and 1800 doubled the price of bread again, and in 1801 the artificers combined in petitions for an increase in the basic rates of pay. The Navy Board would not agree to this, and instead responded in late March 1801 by offering a temporary allowance to married men, varying between 4½d and 1s per day, the amount depending on their occupation and the number of children they had, whilst unmarried men would receive no allowance. At Plymouth, the offer was spurned and some men became involved in food riots. At Portsmouth, the men voted to accept the offer but delegates from the other yards persuaded the committee of artificers to reject it, to conform with the decisions of the committees at the other five home yards. In May 1801, a committee of the Navy Board toured the six yards, on the orders of the Admiralty, who were concerned that the ringleaders had revolutionary intent, and 340 men were discharged. The allowances were paid and the men received no other concessions at this point: the Admiralty did not feel the need for further concessions because the yards were fully manned.

After the brief Peace of Amiens, the war resumed in 1803 and the yards needed more men, putting the Admiralty in a weaker bargaining position. From 1803, men working by the job (on repair work) had their earnings ceiling completely lifted, receiving as much pay as they could make, and in 1805 shipwrights working by task (on new construction) had their articles of work increased in value by 20–25 per cent.[10] Earnings in the dockyard were traditionally paid at least three months after the quarter to which they referred had elapsed. This caused considerable hardship and men had to resort to money lenders at rates of interest of 10–15 per cent, and for security had to pay a clerk in the office of the Clerk of the Cheque a fee for a note confirming they would receive their wages. The situation was alleviated from 1805 when men were paid three-quarters of their usual earnings weekly. Some continued to fall into debt, and following the establishment of the Master Measurer's department, and the introduction of tables which simplified and expedited the calculation of piecework earnings from the work performed, complete weekly earnings (including piecework) could be calculated more quickly so weekly payment in full was introduced in July 1813.[11]

When the Napoleonic Wars ended, many men were kept on to avoid adding to the swelling social unrest and economic depression, so the yard was overmanned. However, because the Navy Board regarded the time wage as a recipe for idleness, the men were kept on piecework, but told to slow down, which they did not do, so the Board resorted to what was still called piecework but was actually time work with continuous check measurement. This policy was overturned in 1822 and deep cuts were made to manning levels, with about one-third of the workforce being lost by 1831. It was the unskilled labourers who bore the brunt of the cuts, so

The Pay Office, built c.1800, was managed by Charles Dickens' father, John Dickens (1785–1851) from Christmas 1807 to January 1815. It is sited to the north of the Porter's Garden.

that surplus craftsmen, who might be needed in the future, were retained. The rub was that they had to undertake the unskilled labouring work, on a rota basis, which they greatly resented. Job piecework, i.e. that on repairs, was notoriously difficult to measure, and the measurer often had to take the shipwright's word for what work had been done, which led to exaggerated claims and intimidation of the measurers. Given this problem, and the fact that even after the cuts the dockyards were still overmanned, the piecework system was ended in 1833 and time work was restored. However, wage classification was introduced as a supposed incentive to work harder. The shipwrights were divided into three classes and paid accordingly: 4s 6d for the first class (the best men, but limited to 20 per cent of the total), 4s for the second class (the great majority) and 3s 6d for the third class (men who had shown 'great insufficiency, negligence or misconduct'). A gang consisted of all three classes, but the work was mostly collective, so the second- and third-class men objected strongly to the system, which nevertheless lasted for a decade.[12]

In 1833, the binary division of the workforce into *established* and *hired* men was introduced. Previously there was no establishment of dockyardmen entitled to permanent employment, although in practice dockyard-trained artificers were employed for life, and men hired from outside the dockyards might be too if they acquired sufficient seniority to survive the culls of hired men that usually occurred at the end of wars. In 1814, superannuation had been extended to all workmen with at least twenty years' unbroken service who were discharged by the Admiralty. This, and the retention of almost all the workforce after the Napoleonic Wars, had encouraged the belief amongst the men that no one would be discharged until he was entitled to a pension. It also meant that the Admiralty had over time to pay an enormous pensions bill. To address this situation, the Navy Board created a permanent *establishment* limited to the number of men needed in peacetime. When more men were needed, they were to be *hired* only for one year at a time.[13]

The onset of the Crimean War in 1854 increased the workload on the dockyards, and there was an imperative to recruit and retain dockyardmen, so piecework was reintroduced for the duration of the war. The problems of measurement difficulties and abuse of recording persisted, whilst the earnings of a shipwright rose from

The entrance to Portsmouth Harbour.
(Engraving by William Miller, after J.M.W. Turner)

4s on day work to 6s per day. When the war ended, piecework was discontinued and day work was re-established, but the wage had to be raised to 4s 6d per day.

Analysing Dockyard Performance

In 1749, Lord Sandwich initiated visitations to the dockyards by the Lords of the Admiralty (something that the Navy Board had never done) because 'the works are not carried out with the expedition that might be expected from them, which must arise from the remissness of the officers, or insufficiency of the workmen, or both'.[14] The alleged inefficiencies of the dockyards have frequently been analysed by dockyard historians. Haas[15] asserted that, 'the dockyards were not well managed … That was the conclusion reached by almost every official enquiry, and the opinion of most First Lords of the Admiralty, between the late eighteenth century and the

early twentieth century.' One reason, he said, was the assumption ('widely held in naval and dockyard circles') that, 'the royal yards built to a higher standard [than merchant yards], even though contracts were performed under continuous Admiralty oversight'.[16] The inefficiencies of dockyards are sometimes explained by saying that they were unique in scale and capacity, whereas private shipyards were far smaller and only built to contract, not carrying out constant repairs; and, said Coats, critics of the dockyards 'paid scant attention to the subjective values, ambitions and alternative aims of the people within the dockyards'.[17] It has been said that the dockyardmen worked hard in times of war but not during peacetime, when the workload was smaller but men were being kept in employment to maintain a capability that would be needed in wartime: Coats[18] contended that the 'loyalty of the labour force and its willingness to work at full stretch when required excused some slowness at other times'. Pollard[19] cited Haas' view that 'repeated attempts at reform and improvement … were blunted and brought to nought by the inertia and obstructionism of the whole organisation', but, he asked, 'Can the world's first navy really have tolerated over such a long period a wholly incompetent organisation to build most of its ships and repair and maintain all of them?'

Other dockyard historians have made similar points. Knight[20] argued:

> Whatever charges of inefficiency can be laid at the dockyards in the eighteenth century, in the final analysis and over the whole period they created in part and were wholly responsible for maintaining an overwhelmingly successful fleet; and any judgement on the administrators, yard officers and the workforce must be seen in the context of British state and society, and if one goes further, comparison with the greatest antagonist, France.

The French yards used impressed labour and convicts, which 'eroded motivation and energy to a far greater degree than in England'. Morriss[21] mentioned, in defence of the dockyards, 'the great burden of work [the dockyards] carried, their achievements in the face of enormous logistical problems, and their share of responsibility in maintaining the British fleet against the attrition of the enemy and the elements'.

But were these valid reasons to excuse the apparent inefficiencies? Samuel Bentham defined 'three sovereign principles of sound management … clear lines of authority, individual responsibility, and full and accurate information about costs, operations and materiel', but Haas conceded that to apply these principles 'would have required a quantum leap', being 'perhaps beyond the knowledge and ability of [Bentham's] day to put into practice'.[22] The Navy Board could exhort dockyard officers to follow its instructions, but the prevailing organisational culture was not conducive to this: Coats notes that there was no lack of instructions, but added that:

> comparing theoretical instructions with the actual behaviour of officers and men reveals the impossibility of a model succeeding without adequate enforcement mechanisms and the near impossibility of imposing an executive ideology upon both a middle management and the workforce.[23]

Dockyard historians sometimes confuse terminology when discussing the performance of dockyards, which obfuscates their analysis. In particular, they confuse *effectiveness* with *efficiency*. Haas pointed out that in view of the considerable expenditure on them, there was a desire that the dockyards should be managed well and give *value for money*.[24] Value for money is a phrase used today by government and in auditing. The National Audit Office[25] provides a definition: 'Good value for money is the optimal use of resources to achieve the intended outcomes. "Optimal" means "the most desirable possible given expressed or implied restrictions or constraints".' Value for money has three components: economy – buying inputs of a given quality at the lowest cost; efficiency – ensuring that the maximum amount of output is achieved from an operation for the minimum amount of input; and effectiveness – ensuring that the outputs of an organisation are as closely aligned as possible to its objectives.[26]

Of these three, we have seen evidence that it was probably efficiency that the dockyards had the most problems with: for example, dockyardmen frequently worked slowly and took excessive breaks, and through their abuse of the chips privilege, wastage of materials could sometimes be high. Thus productivity, a measure of efficiency defined as the ratio of outputs to inputs, could be low. On the other hand, if, as writers such as Knight and Morriss contend, the dockyards met the needs of the fleet – for example by producing good workmanship that stood the tests of time and sea service, and completed work in a timely manner so that the ships were available to the fleet without unnecessary delay – they would rate more highly on effectiveness. So it is probably the case that the dockyards gave quite good support to the fleet, but as this was not always achieved with optimal use of resources, they were inefficient.

Build me straight, O worthy Master!
Stanch and strong, a goodly vessel,
That shall laugh at all disaster,
And with wave and whirlwind wrestle!

Henry Wadsworth Longfellow (1807–1882),
'The Building of the Ship'

Sourcing the Timber

Not for nothing were the ships of the line called Britain's wooden walls: their massive timber construction presented a formidable obstacle to any enemy seeking to invade the country and gave them an advantage in battle over their French adversaries, which were more lightly built. The principal building material was oak, and English oak was preferred, because it was thought to be superior to foreign oak, being slow growing, which gave it a close grain, making it tough and durable, and it was considered to be more resistant to dry rot. A seventy-four-gun Third Rate consumed about 3,500 loads of oak (a load was 50 cu.ft of wood, approximately equal to one large, or 'standard', oak tree). When oak was imported, the preference was to use it for planks rather than frames: in 1760, only 345 loads of foreign oak were imported into Britain, compared with 4,105 loads of plank from the Baltic alone.[1] The English oak was sourced from the royal forests – the New Forest, in Hampshire, and the Forest of Dean were, by the second half of the eighteenth century, the only two still supplying substantial quantities of naval timber[2] – and from private landowners and timber merchants, whose oak was grown in Kent, Sussex, Surrey and Hampshire. For centuries, a plentiful supply of oak was a strategic necessity for the country, but the government did not always do enough to encourage new planting and good management of the forests and copses. In the reign of George I, statutes were passed to preserve all domestic woodlands from destruction. However, opportunities were missed in the royal forests during the eighteenth century: few enclosures for new planting were made, seedlings were not protected from destruction by animals, and young trees were not thinned out. Furthermore, oak plantings on private estates (where the bulk of the timber in the Georgian era would come from) were insufficient, possibly because of a lack of a price incentive, since the price the Navy paid for oak timber was not increasing.[3]

An oak tree took from eighty to 150 years to grow to maturity, and if an adequate supply of timber suitable for shipbuilding was to be available, it was important that trees were not cut down prematurely, as they might be for house-building or charcoal-making, for example. This especially applied to the 'great timbers' (of more than 24ft length) required for stems and sternposts. Since private landowners could not always be relied upon to wait for their oaks to fully mature, and were aware that if they left them too long they might start to decay, the New Forest was used as a source for these great timbers. Acts of Parliament in 1698 and 1808 created the New Forest enclosures, and 6,000 acres of fenced areas were specially planted with oaks and other trees for production of timber for shipbuilding in the Navy's dockyards. An annual quota of felling 300 timber oaks, fifty 'old' (and possibly decaying) oaks and 100 beech trees in the New Forest was established in 1707 to help husband this resource, with the aim of preserving younger trees for the future. Despite being a royal forest, the Navy did not always find it easy to get the trees that the Portsmouth's Yard Purveyor selected because the holders of ancient offices in the forest were able to influence the choice of trees felled, often choosing trees which might be decayed. By 1771, the New Forest supplied about 870 loads of oak to Portsmouth Dockyard every year, out of an annual peacetime consumption of 6,000 loads.[4]

From the time of Cromwell, when large numbers of trees were felled for shipbuilding, until the Seven Years' War, the dockyards were usually able to rely on supplies of English oak being adequate, although there were sometimes shortages and alarms. But the number of trees felled during the Seven Years' War and American War of Independence depleted stocks, which were not being adequately replenished, and during the Revolutionary and Napoleonic Wars, when demand for shipbuilding was again high, it was necessary to import considerable quantities of oak from the Baltic, notably via Danzig, Memel and Riga. The timber shortage was most acute in 1803–05, when it was necessary to use unseasoned timber on some occasions, and to repair ships by doubling the sides of their hulls, using elm and beech in combination with fir and oak, a mixture that would previously been unacceptable. Furthermore, in 1804, the Navy ordered five frigates to be built of fir, one of them, the *Alexandria*, being launched at Portsmouth in February 1806.[5] The extent of use of foreign oak is illustrated by figures for 1812, when Portsmouth Dockyard alone expended 9,253 loads of foreign oak, in addition to 6,231 loads of English oak. Of the oak on order for Portsmouth at the time of the 1813 Admiralty visitation, 13,512 loads were from foreign suppliers and only 2,488 from English sources.[6] This can be contrasted with the situation at the time of the June 1775 visitation, when 16,247 loads of English oak were in stock, together with only 350 loads of foreign oak.[7]

After felling, the timber was transported over land by a team of horses pulling a cart known as a tugg, the simplest of which was two-wheeled with an axle and curved bolster, below which the logs were slung at the leading end by chains after the cart had been pulled into position above them, whilst the other end was dragged across the track. Some tuggs were four-wheeled wagons and carried the load above the chassis. Such transportation was a slow and expensive process, so wherever possible the load was transported by water – perhaps floated down the upper reaches of a river to a point where it could be loaded onto a barge. For this reason the timber was not usually felled at distances further than about 30 miles from a waterway. Hampshire's many rivers, such as the Lymington, Beaulieu, Itchen, Test and Meon, allowed timber to be transported by water via the Solent to Portsmouth fairly easily.[8]

The shape of oak was important: straight timber was usually plentiful and could be used for beams, sternposts, keelsons and planks, but there was a large demand for compass timber, which either had a gentle curve for use in futtocks, or was L-shaped for

Straight, L-shaped and curved timbers would be selected from trees of the appropriate shape. (From the French *Encylopédie Méthodique Marine*)

knees (which joined the transverse beams to the frames). In dense groves of trees, the oaks grew straight and tall as they reached for the light; in hedgerows and on the edges of small copses, the oaks could take more spreading shapes and were the source of compass timber. In the dockyard, the oak had to be left to season for two to three years before use, or if the timber had been shaped whilst still green (which made it easier to work), it could be left to season with the ship in frame. Until the 1770s, Portsmouth, like the other English dockyards, did not have timber seasoning sheds, and ships were not usually built under cover, so there were always problems with decay. In the middle of the eighteenth century, the life of a hull was about twenty years, but as the century wore on, this figure reduced to an all-time low of eight years by 1792. Many ships had to be almost completely rebuilt during their lifetime, in what were called large repairs.[9] If timber was in short supply, ships were sometimes built and sent to sea with green oak, often with disastrous results from premature rotting of their frames, typically halving their life. For best results, oak was usually felled in the winter, when the sap was not rising, because such timber was believed to be more resistant to rot than that felled in the spring or summer. The oak for frames was sometimes charred, which was thought to increase its life, and snail crept – which entailed gouging out crooked channels in the surfaces of timber to gain circulation to parts where air would otherwise not circulate. The *Royal William*, launched at Portsmouth in 1719, had a very long life, and was finally dismantled at Portsmouth in 1813, when the opportunity was taken to examine her timbers for clues as to her longevity. It was concluded that the ship owed its long life to well-seasoned timber and careful workmanship, and not to any special procedures such as charring and snail creeping.[10]

To increase the life of ships, the Navy Board sought to improve its storage by using covered seasoning sheds, stacking the timber in a way that allowed air to circulate around it, and aiming to increase timber stocks to equate to three or four years' supply, so that it might be well seasoned when used, but was usually unable to achieve this level of stock. Timber seasoning sheds were built at Portsmouth in the mid 1770s, the last being completed in 1775.[11] In 1813, Portsmouth had nearly two years' supply of English oak and one and a half years' supply of foreign oak, which was more than there had been for many years. However, there was a shortage of 'some pieces

Sawyers at work, with the hands of the bottom man in the pit just visible. (From *The Book of English Trades*, 7th edition, 1818)

of particular shape and dimension', which meant that although the timber stock would be sufficient to build the greater part of four or five seventy-four-gun ships, it was probable that a single one could not be completed, except by scarphing pieces together.[12]

English elm was another important material, having a great resistance to rot when fully immersed in water, and was used for the ships' keels and garboard planks – the three planks above the keel on each side of the ship. Fir (such as Scots pine) was used for masts and spars, and was sourced from Norway, the Baltic and the

American colonies. In the reigns of William and Mary, and George I and II, statutes were passed which aimed to preserve white pines in New England for large lower masts.[13] Fir was also used to clad the underside of hulls to protect them from marine borers, until the use of copper sheathing was introduced. In the early nineteenth century, larch, first used in 1792, was recognised as a suitable alternative to oak for planking.[14]

The timber received in the dockyard was usually in its raw state, unshaped and with its bark still on, though some of the imported timber had been 'sided' – with flat surfaces on two opposing sides – or converted into planks, to reduce transportation costs. The sawyers in the dockyard sided unconverted timber and, when required, cut it into planks or straights of reduced thickness. They worked in pairs at a saw pit, with one sawyer (the 'top dog') standing at ground level, holding the top of the vertical saw, and another (the 'underdog') standing below the log, in the pit. The demands of the dockyard meant there were numerous sawyers: for example, 134 were employed at Portsmouth in 1774 and 241 in 1813.

The Work of the Shipwright

Before looking at how the ship was constructed, we will consider some aspects of the shipwright's work, since these were central to understanding how the ship was built; and the shipwright was easily the most numerous, and arguably the most important, of the trades employed in the dockyard: for example, 861 shipwrights were employed at Portsmouth in 1774, and 931 in 1813. The shipwright's training included not only the main skills of ship construction and repair, but also more specialist ones such as caulking, work in the mould loft, boat-building and mast making, and a few shipwrights were accomplished wood carvers. Some shipwrights chose to specialise in boat-building, mast making or 'lofting', but caulking was normally a separate trade.

The shipwright's best-known tool was the adze, similar to an axe but with the blade set at right angles to the handle, not in line with it. Using an adze was known as 'dubbing', and included both coarse preparatory work in shaping timbers and the most delicate operations in finishing them. Equally good on either the flat or the round, when sharpened to a razor's edge it could finish a plank to almost the same standard achieved with a joiner's plane. Axes, hand saws, large chisels, gouges and mauls (hammers) were also important tools, together with augers for drilling, wooden braces and bits, and spoke shaves amongst others.[15]

As well as shaping timber, the shipwright had to be able to join pieces together, known as 'faying'. Futtocks might be joined by angular pieces called chocks located into triangular cut-outs at the ends of the two futtocks to be joined, with bolts securing each side of the joint. For fairly straight timbers, such as beams, the keel, keelson and deadwood, overlapping joints, set at an oblique angle and known as scarphs, were used. A variety of scarphs was used in different applications. They often had steps in the joint, or were drilled to receive a 'coak' (dowel), to prevent sliding, and were reinforced with bolts. Where stepped, wedges might be driven in from each side at the step to make it tighter. Hook and butt scarphs were formed with a hook or projection, for one part to form into the other, as used in the clamps. For pieces meeting approximately at right angles, a mortice and tenon joint was used, i.e. with a projecting tongue locating into the female mortice, for example where the sternpost fitted into the keel; if the join was in the horizontal plane, a dovetail joint might be used, for example in the beams, carlines and ledges.

To fix the timbers in place during construction of the ship, a number of different fastenings could be used. Bolts were used to fix timbers into place in much of the internal structure, including the frames and knees in the hold. The longest bolts through the deadwood might be about 17ft long, and the upper bolts through the knee of the head 19ft. To prevent the bolt from drawing, the point (end) was clenched, that is spread or riveted on a ring; alternatively it might be 'forelocked', with a thin circular or straight wedge of metal passing through a mortice at the point. Bolts were made of iron or, if below the waterline, 'mixed metal', a copper/zinc/tin alloy. Wooden pegs known as trenails (or treenails, and pronounced trunnels) were used to fix most of the rest of the structure together, including their main application, the planking of the hull. They had the advantages that they did not rust and did not damage the shipwright's adze if accidentally struck. In a Third Rate ship, they would be 1½in in diameter. Sometimes the trenail had a wedge hammered into a groove across the diameter of the head to give a tighter fit. Iron or mixed metal nails (known as spikes) were used for deck planks and sides of the parts to be joined, and had jagged edges on the two parts that were driven in. The join was purposely made quite weak so that the false keel could break away quite easily if the ship ran aground, without causing damage to the main keel.[17]

The names for parts of the ship were often intriguing – deadwood, knees, fashion pieces, gripe, spirketting, futtocks, buttocks and nogs, to name just a few. These names and the methods used in construction of the ship varied – sometimes confusingly – and hence it is often necessary to give the synonyms when describing the parts and methods.

Constructing the Hull

John Fincham, Superintendent of the School of Naval Architecture in Portsmouth Dockyard, gave a description of the parts and methods of construction of a wooden sailing ship-of-the-line in *An Introductory Outline of the Practice of Shipbuilding, Etc (1825)*. He summarised the principal timbers that formed the structure of the ship:

> The longitudinal form is determined by timbers called the keel, stem, and stern post. The stem, which is at the foremost extremity, is supported in its combination with the keel, which is the lowest part of the structure, by other timbers lying in its concave part, called the apron and stemson: the apron and stemson unite with timbers called the deadwood and keelson, which timbers give support to the keel: the stern-post, which is at the aftermost extremity, is supported by timbers called the inner stern post and sternson; and these timbers likewise form a junction with the keelson, deadwood and keel, so that a mutual connexion is kept up by them to preserve the longitudinal form.
>
> Transversely, form is given and an [*sic*] union formed by assemblages of timbers placed vertically, extending equally on each side called frames. The lowest timbers of the frames, called floors are connected with the deadwood and lie between the keel and keelson, extending equally on each side; the other timbers of the frames, called futtocks and top timbers keep up a connexion from the keel to the timbers that form the upper boundary of the structure, called gunwales and plank sheers.

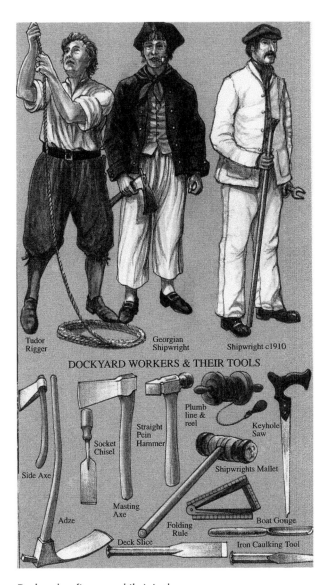

Dockyard craftsmen and their tools.

for the ribbands which held the frames in position until the planking was laid, but not for structural applications.[16] Staples were also used in a few applications: they were bent fastenings of metal, formed as a loop, and hammered in at both ends. The false keel was fastened to the main keel with copper staples, which were driven in from the

The frames: 'Showing the manner of forming the frame on the old and new principle with the mode of securing it before the beams are in place, and the exterior plank wrought.'
(From John Fincham, *An Introductory Outline of the Practice of Shipbuilding, etc*, 1825)

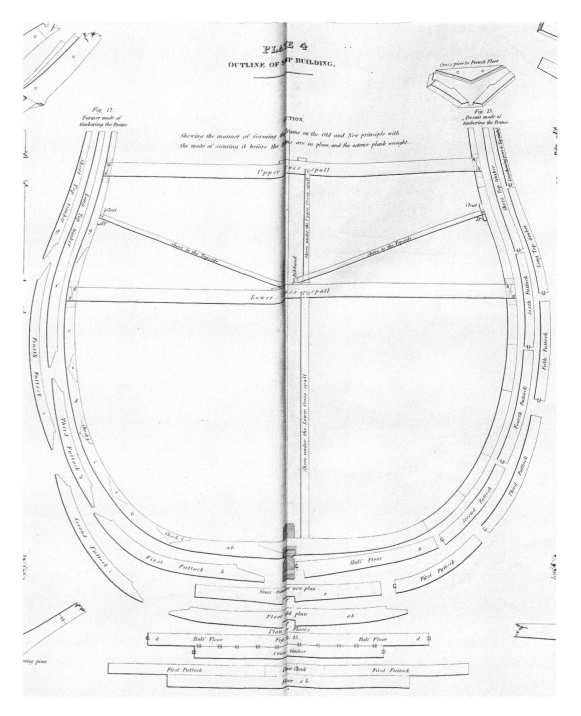

PLATE 4
OUTLINE OF SHIP BUILDING.

The longitudinal form is further maintained and connexion preserved, by exterior and interior linings called planking, and the interior binders, called shelf-pieces, united to the frames. The exterior lining or planking which is connected with, and covers the whole surface of the frame, is made water-tight, to preserve the buoyancy of the body. The two sides are connected and sustained at their proper distance apart by timbers lying horizontally, called beams, firmly united to the sides of the ship. Platforms, called decks, are laid on the beams, on which the guns and other necessary furniture of the ship are placed; and which afford cabins for the accommodation of the officers and ship's company.[18]

The paper plans of the ship, 1/48 scale, were received by the dockyard and sent to the Mould Loft, where full-sized lines of each curved part, including the frames, were inscribed on the floor by shipwrights (who were known as loftsmen). The spacious planed deal floor was covered with black plaster known as gesso, and the loftsman, who wore canvas shoes or went barefoot to avoid damaging the floor, drew the lines in chalk. Joiners then made moulds (templates) from fir or deal about ¾in thick to represent the inner and outer edges of the frames, etc.[19] These were taken down to the timber stacks and the building slip, where they were used to select the frame pieces. First the sawyers would 'side' the timber, i.e. cut flat surfaces on two opposite sides: then the shipwright would 'mould' it, using a variety of tools including an adze, an axe, chisels, a set of augers for drilling and a small saw, thus shaping it to match the pattern received from the loft. The complete timber could require scarphing together two or more pieces, or joining them with tenon or dovetail joints.

The building slip was inclined at an angle of about four degrees towards the water, and along its centre-line were placed large hard, knotty oak blocks upon which the splitting blocks were placed. These were clear-grained timber with no knots, which would be cleaved out cleanly when the false keel was put on and the ship was ready for launch. The false keel gave the keel some protection from damage if the ship ran aground, and from boring shipworms; it also gave extra depth to the hull, which helped prevent the ship from making leeway when sailing close-hauled or with the wind on the beam. The top of it was coated with a tar and hair mixture to deter the worm from penetrating the main keel. The first stage of construction was the laying of the elm keel on the blocks in sections, positioned fair and straight by eye and scarphed to one another, with eight bolts through the scarph. The sections were 'nogged' with short trenails (known as nogs) driven diagonally from the sides of the keel into the blocks, to keep it in place while building. A rabbet (a rebate or groove) was cut all the way along the upper edge of the keel on each side, to receive the edge of the lowest plank, the garboard strake.[20] Frequently, a temporary fir keel was laid first, upon which the main keel was placed: the fir keel protected the main keel during building, as it was found that when ships stood for a long time on the slip (including the time to allow the frame to season) the permanent keel became decayed.[21]

At each end of the slip was a pair of sheerlegs, large straight timbers supported by rope guys and angled towards each other to meet at their heads, where they were lashed together: they were fitted with block and tackle for lifting the stem and stern posts into place. A rope ran between the two pairs of sheer legs, from which tackles could lift frame timbers into place. The stem (or stem-post) was made of two or three pieces and was fitted into the end of the keel with a complex scarph, known as the boxing (or boxen) scarph, and curved sharply up from it. It also had a rabbet to receive plank ends. Inboard of the stem-post was a false stem-post, or apron, to reinforce it: this was bolted to the stem through the scarphs in the latter, and the assembly was then raised into place and supported by shores, which were nogged into the stem to provide temporary security. To support the bowsprit, vertical timbers called knight-heads were bolted to the sides of the stem or apron and separated by stem-pieces to allow the bowsprit to pass between them. The sternpost, which tapered towards its head, was tenoned into a mortice in the after end of the keel (with two tenons) and had a rabbet cut into it all the way up it to accommodate the ends of the planking. A false (or inner) sternpost was bolted to the inboard side of the sternpost and had grooves to receive the transoms, which were transverse timbers that helped form the stern and support stern galleries and deck. The vertical part of the stern assembly consisted of the fashion pieces, of complicated shape, curving outwards and backwards, which were connected to the transoms and the deadwood near the end of the keel. By the middle of the eighteenth century, it was common for all of the stern structure (stern-post, inner post, transoms and fashion pieces) to be assembled on the dockside and raised as a unit to be

The stem and adjoining timbers. (From John Fincham, *An Introductory Outline of the Practice of Shipbuilding, etc*, 1825)

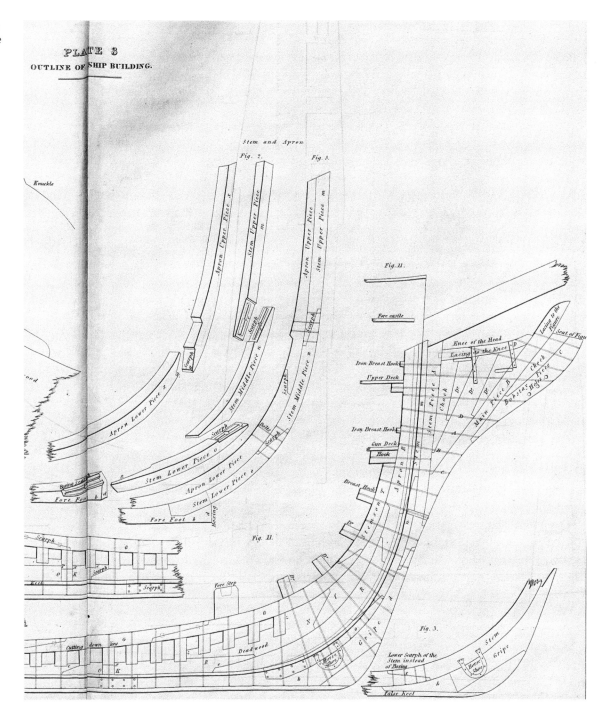

supported by shores. Within the basic structure of the apron, keel and inner sternpost, the deadwood was fitted longitudinally on top of the keel to help strengthen that basic structure and give the seating of the floors. The fore and aft deadwoods were like massive knees binding the stem and stern posts to the keel, providing a solid foundation for the half-timbers and the plank ends where the hull narrowed towards the bow and stern. The central deadwood ran the entire length of the keel and received the floor timbers, which fitted into larges grooves cut into it.[22]

Each frame was made up of several sections, known as floor timbers, futtocks and top timbers, and the frames were arranged in pairs, side by side, either touching or slightly separated, with the joins between sections in one frame displaced from those of the adjacent frame. Each floor timber spanned the deadwood (which was let into them) to form the lowest part of a frame, whilst the futtocks and top timbers above it were made in pairs so that one side of the ship was the mirror image of the other. The frames were placed in pairs, side by side, and arranged so that the joins in one of them were as far as possible from those of the other. The two adjacent frame bends in a pair were bolted together, with a chock in the space between them if they were not touching each other. To allow for gun-ports, some of the frames were 'filling frames' rather than complete frames. Each pair of frames was close to the next, so that the structure was about two-thirds solid under the planking. Whilst the frames in the central part of the ship were vertical, towards the bow and stern they were angled away from the vertical and known as cant frames. Such frames had half-timbers rather than floor timbers: these were made in two halves which joined above the keel by being butted into the deadwood. The bow of a large ship was very bluff, and close to the stem the cant frames were replaced by vertical timbers known as hawse pieces, into which the hawse holes were cut. The hawse pieces filled the space between the foremost cant frame and the knight-head. In the stern, the fashion pieces, fitted earlier, were actually the last cant frames, whilst the horizontal transoms were augmented by a series of upright timbers, straight or curved, which helped form the shape of the stern and counters.

The assembly of the frames into the ship started by fitting the floor timbers and half timbers. The futtocks of each 'frame bend' were then bolted together and raised into place as a unit, both sides of the frame being raised at the same time, the first being the 'full frames', i.e. those not interrupted by a gun-port – which would form the sides of a gun-port. Then the port cills were fitted to form the top and bottom edges of the gun-port, after which the filling frames could be fitted. When the assembly of the frames was complete, the keelson was placed over the floor timbers, helping to lock them into the deadwood, and was bolted through the deadwood and keel with a single bolt. Then the stemson was bolted onto the apron and scarphed to the fore end of the keelson, and the sternson was bolted against the inner (or false) sternpost and scarphed to the aft end of the keelson: together, the stemson and sternson formed extensions of the keelson. Ahead of the stem, the knee head was fitted: this was an S-shaped combination of timbers which supported the figurehead and to which the ropes securing the bowsprit could be fixed. During construction, before the fitting of the deck beams, the frames had to be supported and kept in position in several ways: ribbands were long, narrow, flexible pieces of fir nailed longitudinally upon the outside of the ribs, and three sets were normally fitted below the maximum breadth on a Third Rate ship; cross spalls were light timbers connecting the two sides of each frame; and shores were placed outside the hull to keep it upright on the slip. Once in frame, the ship was often left for some months – sometimes more than a year – before planking, so that the timbers could season.[23]

The frames were covered on both the outside and inside with planking, mostly fixed to each frame with two trenails, but by the nineteenth century the ends of the planks were fastened with iron or copper bolts, the latter being used below the waterline. The inner planking was known as the ceiling. Many planks were curved to fit the shape of the hull: this was generally achieved by using a steam chest to make them more pliable. The shipwrights carried the hot planks to the ship's side, where each plank was lifted by ropes into its approximate position and then forced into place using 'wrain staves', working in from the bows and the stern. However, thick planks with a sharp curve, such as the wales at the bow, had to be carved to shape. As they were fitted, planks were butted together at their ends to form strakes – which in most cases extended from stem to stern, but near the bow the width of the planks had to reduce and some planks came to an end before they reached the stem-post. The thickest planks were the wales and rails: the wales were the first parts of the planking to be applied, running below the gun-ports at each level and helping provide longitudinal strength to

The parts of the sternpost and adjoining timbers, and a view of an elliptical stern showing the disposition of timbers. (From John Fincham, *An Introductory Outline of the Practice of Shipbuilding*, etc, 1825)

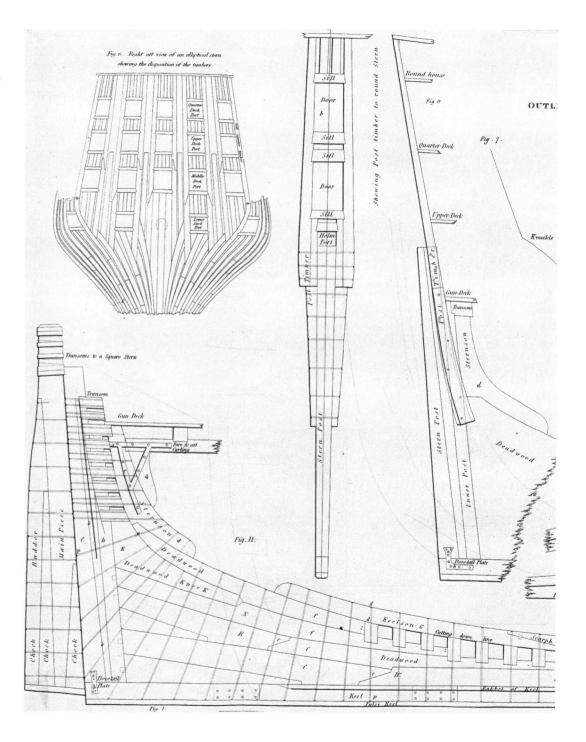

the structure. They had more complex interlocking joints between the planks of a strake, rather than the simple butt joints of most strakes. On a Third Rate, the main wales (the lowest of the wales) were made up of three or four strakes and might be about 8½in thick, compared with the 4in thickness of normal hull planking. Other principal strakes were known as the sheer strakes and defined the curve that forms what is known as the sheer.

Up to eight gangs of men would be engaged in planking, working up and down from the wales, with another gang working upwards from the garboards. The planking of the bottom was a complex task because of the shape of the hull: for example, near the bow the planks had to be curved upwards as well as inwards. Where adjacent planks were of different thicknesses, the thicker one had to be faired in with an adze to produce a smooth transition rather than a step, so that the water flow over it might be smoother. The external bottom planking began under the main wale, and the very bottom plank, which fitted into the rabbet of the keel, was a principal strake known as the garboard strake and was often elm rather than oak. The external side planking was easier to apply because the sides of the hull had comparatively little curve, though many strakes were interrupted by gun-ports. The strakes just above the wales, known as 'thick stuff', were thicker than normal but were not faired in to the adjacent strakes. The sides were topped by a flat timber on the gunwale, fastened over the heads of the timber and the edges of the planking. Before the internal planking was fitted, the clamps were probably fitted first: they were heavy timbers along the insides of the frames, following the normal line of the decks, and had recesses cut in them to receive each deck beam, which they supported. The internal planking of the hold had thicker strakes at the positions which covered the joins in the futtocks: again, this 'thick stuff' was not faired into the adjoining thinner planks, though the corners were usually trimmed off to give a neater appearance. The other internal planking followed the lines of the decks rather than the futtock heads and was usually fitted after the deck beams were in place. Mast steps were fitted over the keelson and internal planking to provide seats for the end of each mast.[24]

The deck beams spanned the width of the hull at each deck level, normally 4ft or 5ft apart, and rested at their ends on the clamps, into which they were let by ¾in. One beam was placed under every port, and one between every two ports, and they were placed to be immediately over one another so that strengthening pillars between them could be positioned directly above one another to receive support at the lowest level by the keelson. The beams were cambered, with their centres higher than their outside edges, except for the orlop deck which was flat. The camber gave structural strength and allowed water shipped in heavy weather to be drained away to the waterways; it also reduced the travel of the guns when they recoiled after being fired. In a ship-of-the-line, the deck beams were mostly too long to be made from a single piece of timber, and in most cases between two and four pieces were scarphed and bolted together. The spacing of some of the deck beams was interrupted by features placed along the centreline of the ship, such as hatchways and masts, so half beams were also used: these started at the side of the ship in the normal way, but curved either forward or aft to join the adjacent full deck beam. At the sides of the ship, the beams were supported by L-shaped knees of three types: hanging knees were placed vertically and supported the beams from below – but the horizontal arm was secured alongside the beam not under it, to save space; lodging knees were placed horizontally in the angle between the beam and the side of the ship; and standards were secured vertically above the beams. The knees were normally made of oak, but iron knees were later often used in the latter part of the wooden ship era, and the arms of the knees were bolted to the deck beams and the sides of the ship. (Before the use of iron knees, when knee timber was in short supply other solutions to form this vital union had to be found, such as combining wooden chocks under the beam, bolted through the side, with iron plate knees fixed on each side of the chock and beam.) Additional support was given to the beams by the pillars, which were mostly fitted at or near the centre of each beam.

At the bow, a system of 'breast hooks' was used for internal strengthening, whilst at the stern, 'crutches' and transom knees were similarly employed. The interior of the hull was further strengthened by 'riders', which were essentially interior ribs or frames fitted over the ceiling. However, Robert Seppings, master shipwright at Chatham, devised a system of diagonal riders which was first partially and successfully adopted in the *Kent*, a seventy-four-gun ship, in 1805. These diagonal riders were of wood in ships-of-the line or iron in frigates, were securely fixed to the inside of the ship's hull and produced a far more rigid structure, rendering

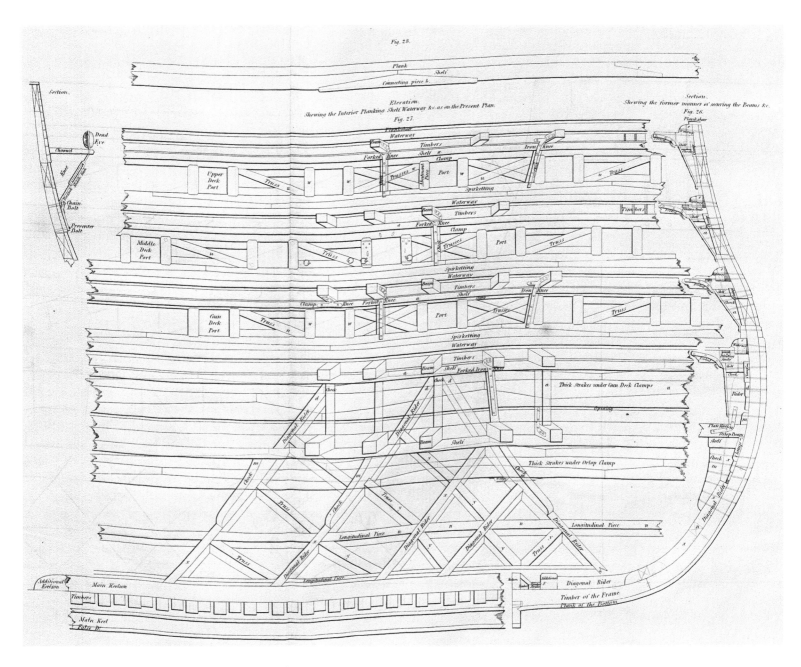

The interior planking, shelf, waterway etc. (From John Fincham, *An Introductory Outline of the Practice of Shipbuilding, etc*, 1825)

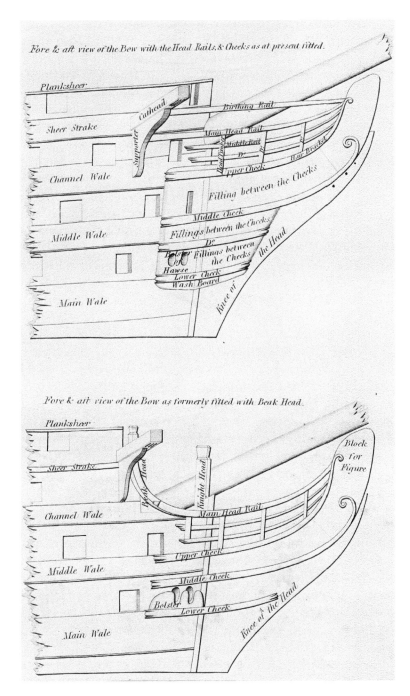

Fore & aft view of the Bow with the Head Rails & Cheeks as at present fitted.

Fore & aft view of the Bow as formerly fitted with Beak Head.

the old type of rider obsolete.[25] Heavy vertical timbers called bitts and catheads were fitted forward on the hull as part of the anchor gear, and smaller bitts were placed around the masts for securing the running rigging. To support the deck between beams, carlines ran fore and aft and were fitted into the deck beams at each end, with three rows on either side of the centreline on a seventy-four-gun ship, and ledges ran transversely across the ship and were dovetailed into the carlines. Before the deck could be laid, the waterways had to be put in – they formed the corner between the edge deck plank and the side of the ship – and the capstans and steps for the masts had to be fitted. The upper deck was the first to be planked, thus affording protection from the weather for work on the interior of the hull. Deck planking was 3in or 4in thick and was fastened to the deck beams and ledges with copper-zinc alloy nails. Danzig deal (Baltic pine) was used for most of the planks, but the outer five or six on each side of the gun decks were of English oak to withstand the wear and tear imposed by the guns and their carriages. The butt joints between planks were offset from those in the adjacent planks in a four-step system; the nails were hammered into pre-cut recesses which were then covered with a small piece of timber to give a smooth surface. For decks below the weather deck, iron nails were used.[26]

The beak, a heavy platform jutting out from the hull just below the bowsprit, was fitted, extending down to the knee of the head. Decorative carvings were added to the hull of the ship, including the figurehead, beakhead and stern decorations, and sometimes decorations on the sides or gunports. Belaying points, cleats and other fittings associated with the running rigging were fitted, as were the rudder stock (which was bolted to the stern-post), rudder blade and tiller. The hull was caulked by caulkers, who drove strands of old rope into the gaps between planks with caulking irons and hammers, and sealed it with hot melted pitch or resin. Above the waterline, a mixture of resin and oil was usually applied to the hull below the level of the main rail, giving a partly transparent finish

Fore and aft views of alternative bow arrangements. (From John Fincham, *An Introductory Outline of the Practice of Shipbuilding, etc*, 1825)

through which the wood colour could be seen. Above the main rail, the hull was painted and sometimes had gilded friezes applied, whilst the figurehead and other carvings were painted in gold leaf or yellow ochre. Joiners (or house carpenters) worked to install the internal fittings of bulkheads, storage compartments, cabins and partitions.[27]

The Launching and Fitting Out

The ship was now ready for launching, on a day chosen by the dockyard officers for its high spring tides. Two wooden rails, known as bilgeways or ways, were built on each side of the keel, and the two halves of a wooden launching cradle were built under both the bow and the stern, to ride on the greased surfaces of the ways. Wedges were driven in under the rails to lift the ship off the splitting blocks, which were then removed, thus transferring the weight of the ship onto the cradles, which were supported by shores. As the blocks were removed, sections of the elm false keel were fastened to the main keel by iron staples, which were driven into the sides of both the keel and the false keel. Before the launching ceremony, large flags were flown from temporary jackstaffs fitted in the mast positions (the masts were not yet fitted) and at the bow and stern. The launching ceremony took place at high tide and may well have been carried out by a dockyard official and, by 1802 at least, it was common for a bottle of wine to be broken over the bows of the ship. To signal the moment of launching, a bell was rung. The ship was restrained by ropes leading from the hawse holes to bollards at the head of the ship, which were then cut, allowing the ship to slide into the water, accompanied by the cheers of onlookers. She was then towed into a dry dock for the hull sheathing to be applied. Below the waterline, the hull was treated with a mixture of hair and tar and then sheathed in light fir boards, to protect against shipworm; the surface was then coated with 'stuff', an anti-fouling composition to deter marine growth. From the 1770s onwards, copper sheathing replaced the fir boards, being more effective as a barrier to shipworm.[28]

The ship might then be laid up in Ordinary (reserve), in which case little further work was done on her. If she was to be fitted out for sea service, then the masts and rigging had to be erected, sails made and fittings such as anchors and their tackle, boats and pumps added. Finally, the guns would need to be shipped. The masts consisted of two or three sections rigged above each other, and were made in the mast houses by shipwrights. If a suitable single length of timber was not available, a section might be comprised of several pieces joined together. The lower masts were fitted by riggers with the ship alongside a sheer hulk in the harbour. The sheer legs on the hulk lifted the mast section to the vertical before it was manoeuvred and lowered into place through the holes in the decks, and its heel was fitted into the tenon over the keelson; the holes were larger than the mast itself, so wedges were driven in to secure it against the 'mast partners' surrounding the hole, and then the standing rigging was set up; the sheer hulk also lifted the bowsprit into position. After the 'top' and its surrounding structure was fitted to the top of the lower mast, the topmast could be hoisted up using a top rope attached to the head of the lower mast, and secured in place. The topgallant mast could then be fitted it a similar way, and the yards and other spars could be raised and rigged. It took two weeks to rig a Third Rate ship.[29]

The sail loft was where the sailmakers worked, and had a smooth wood floor on which the sailcloth was laid out and marked with charcoal lines for cutting out. The canvas sailcloth came in a variety of thicknesses, the choice of which depended on the application. The lightest, known as No. 7, was used for studding sails and set in only the lightest wind, whilst the heaviest, No. 1, was used for the courses and topsails to be set in the heaviest weather, and was over twice the weight of the No. 7. A double thickness of canvas was used in parts of the sail which were most subject to wear. The parts of the sail, once cut, were sewn together by hand using a double seam with between 108 and 116 stitches to the yard. The twine for the larger sails was waxed by hand with beeswax mixed with a sixth part of clear turpentine. Around the edges of each sail, a rope, known as the bolt-rope, was sewn in to strengthen it. At the corners of the sail, loops known as cringles were formed into it; some sails had cringles in their sides for fixing reefing tackles, and buntline cringles in their lower edges to attach the ropes for furling the sail. Many of the square sails had light ropes which passed through the sail running across them, as reefing points.[30] In 1774, fifty-one sailmakers were employed at Portsmouth.[31] Their work also included making canvas covers for boats, dodgers (canvas shelters), cot bottoms, hammocks and tarpaulins.

> Discuss unto me; art thou officer? Or art thou base, common and popular?
>
> William Shakespeare (1564–1616), *Henry V*,
> act 4, scene 1, line 37

The cupola atop the Academy housed an observatory and was also used as a space for teaching fleet tactics.

At the start of the eighteenth century, the training of naval officers was conducted at sea, boys aged between 12 and 14 being taken as volunteers-by-order or captains' servants. The instruction they received from ships' officers was augmented by broader education from ships' chaplains. Such provision was inevitably of variable quality.[1] In 1702, a decision was made to send schoolmasters to sea, with a corps of at least ninety being envisaged, and between May 1702 and August 1705 some sixty-two certificates of competence were awarded to such schoolmasters, though by the early 1720s applications had reduced to single figures in each year. The scheme did not come near reaching the overall complement envisaged, which may have been a factor influencing the decision in February 1729 to establish a shore-side academy, which was to be the first shore-based training establishment in the Navy.[2]

The Navy Board instructed the Portsmouth officers:

> The Right Honourable the Lords Commissioners of the Admiralty, having been pleased ... to direct that an Academy should be erected in the south east part of His Majesty's Yard at Portsmouth ... for the reception and better educating and training up to 40 young gentlemen for His Majesty's service at sea, instead of the establishment now in force for volunteers on board His Majesty's ships; as also for the reception of a Mathematical Master, three Ushers, and a French Master for their instruction, with proper outhouses for accommodation.[3]

Work on the Royal Naval Academy building started in 1729 and was completed in 1733 at a cost of £5,772. Because of its specialist nature, it was not designed at Portsmouth by the dockyard officers, as was normally the case for dockyard buildings, but in London, probably by the Navy Board's Surveyor, Sir Jacob Acworth.[4] It is the oldest building in the dockyard that was not designed by dockyard officers, and its H-plan structure was constructed of grey and plum red bricks (the latter being distinctive of Portsmouth) with Purbeck stone detailing. An attractive original feature of this handsome building is the central cupola, an observatory, though its floor space was used mainly to teach fleet tactics using small fleets of ship models that could be overlooked by the scholars from the timber gallery above.[5]

The dockyard Commissioner oversaw the Academy as its governor and was awarded an allowance of £100 per annum for 'the pain and trouble it will give him to discharge that office'.[6] He had to go to the Academy frequently to inspect and report on it.[7] The wages and salaries, per annum, of the original teaching staff (some of whom clearly had other duties and salaries in addition) were set out as follows[8] and totalled £608 10s including the Commissioner:

Head Mathematical Master	£150
Second Mathematical Master who is likewise to teach writing and arithmetic	£100
Teacher of drawing and the French language	£100
Teacher of fence and dance	£80
A person to keep the arms clean	£10
Surgeon of the yard for physic and attendance	£20
Master Attendant: 10/- for each lesson, 26 lessons in the year	£13
Master Shipwright: 10/- for each lesson, 26 lessons in the year	£13
Boatswain of the Yard: 5/- per lesson, 12 per year	£3
Gunner of Ship in Ordinary: 5/- per lesson, 12 per year	£3
Person to teach the exercise of the firelock: 5/- per lesson, 12 per year	£6 10s
Contingent Charges	£10

The Academy was to board and teach forty scholars, who were to be the sons of noblemen and gentlemen, aged between 13 and 16 on admission. Entrants had to have already achieved 'some considerable proficiency in the Latin tongue'. The senior mathematical master was to be the head of the school, assisted by three 'ushers' (as above) who would teach 'writing, arithmetic, drawing, fortification and all other parts of the mathematics'. Fencing and the 'exercise of the flintlock' would not be introduced to scholars until after their first year in the Academy. In addition to the academic subjects, students in their second year worked twice a week in the dockyard, under the direction and instruction of the Master Attendant, Master Shipwright and Boatswain, hence their inclusion in the table of salaries above. One of the ships in Ordinary would be placed near the dockyard so that the scholars could rig and unrig her frequently, and two guns would be mounted with powder and shot so that they could be instructed in the use of cannon by an experienced gunner from the ships in Ordinary. Each scholar would be provided with a suit of blue clothes yearly, at their own expense, on the king's birthday. They would be entered as Volunteers by Order and receive able seamen's pay, though having 'the privilege of walking on the quarterdeck'.[9]

The broad academic and practical nature of the syllabus stood in marked contrast, not only to the narrow classical curriculum of the eighteenth-century public school, but also to the uneven tuition provided afloat. Each boy had his own room, something that was uncommon at the contemporary public schools.[10] From 1749 onwards, the curriculum also included Latin, for which a master was appointed. Once they had completed their two to three years' studies at the Academy, the scholars had to serve two years at sea before they might be rated as midshipmen. However, they needed to serve only four years' qualifying time before the lieutenants' exam, whereas all others must have served at least six years at sea, as an incentive to follow the Academy route rather than proceed straight to sea. However, the college was not welcomed by naval officers because it undermined their own privilege of patronage in appointing boys to their ships, and this attitude was to dog the Academy for most of its existence. Such officers usually felt that the best education was the culture of a seagoing life.[11] Furthermore, scholars at the Academy did not have the same opportunities as their peers in ships to make the social contacts that could benefit their careers. But the alternative to Academy attendance, instruction at sea, was hampered by the 'unintellectual mediocrity' of an average wardroom made up of 'snorers, chessmen and soakers' who read nothing but the Navy List, and the boys received more gin than instruction, or passed the time 'sleeping, playing, and walking the decks with their hands in their pockets'. Some captains, especially those in frigates, did take more trouble in providing the resources and environment for the education of their charges, but they were far from a majority.[12]

Overall, the quality of both headmasters and staff at the Royal Naval Academy over seventy years seems to have been quite high. The five headmasters were eminent scholars, with one possible exception because not much is known of him. And the Academy seemed mostly to be able to attract and retain high quality staff. The 1749 visitation by the Admiralty Board noted that, 'they had examined into the conduct and abilities of the Masters' and had concluded that, 'great care was being taken of the young gentlemen in every part of their education'. A 1771 enquiry repeated this view, confirming that 'the masters are thoroughly qualified in their respective stations'. One naval officer, a distinguished frigate captain who was to rise to the rank of admiral, Byam Martin, concluded that the teaching staff 'were excellent each in his particular branch', the second mathematical master was 'excellent' and the French master 'first rate'.[13]

But there were some disagreements between staff: in 1735, John Bellonet, French master, made a disposition before justices that his colleague, John Ham, had tried to kill him. The third headmaster of the Academy, John Robertson, seems to have had a poor relationship with his second master, Mr Waddington.[14] Robertson had, in June 1766, complained to the Commissioner about Waddington's family, saying that the latter's 'wife and servant never let a week pass without uttering their illiberal abuse notwithstanding they never have any reply', and that he had had 'opprobrious language vented to my face, my children and wife have been rudely shoved and called by the vilest of names … this abominable spirit is even exerted in the place of divine worship'. Moreover, Mrs Waddington, 'to try how far she could push her malevolence did … violently thrust her elbow against Mrs Robertson's breast saying at the same time "take that you bitch"'.[15] Whatever were the full facts of the dispute, the Commissioner took a hard line and both Robertson and Waddington were dismissed.[16]

Enrolments at the Academy apparently got off to a slow start, with only a handful of pupils present by February 1734, though it began to fill up rapidly after that. However, the 1771 Admiralty visitation found that, 'the number of scholars is never complete and that at this time consists of no more than 14', but stated that the upper limit for pupils was thirty (and not the forty originally specified.) They attributed the shortfall not to inefficiency or unpopularity, but to high fees – noting that, 'the charge of maintaining a young gentleman in this foundation is equal to that of Eton or Westminster.' After an Admiralty visitation in 1773 and the subsequent provision for sons of sea officers to be admitted at public expense, numbers began to rise, and there were at least forty pupils present by 1775, increasing to forty-five by 1800 and fifty-six by 1803. But the free places for the sons of sea officers that were taken up seem only to have attracted the sons of their widows.[17] Their annual fees of £25 for Board and £5 for uniform were placed on the Ordinary Estimate.[18] By then, the curriculum included writing, mathematics, navigation, gunnery, fortifications, dancing and the use of the firelock, and scholars could stay for up to five years.

The conduct of the Academy and behaviour of the pupils have been portrayed by some commentators as one of lax discipline, with a student body characterised by swearing, drunkenness, idleness and riotous assembly. The evidence cited for this usually comes from Dockyard Commissioners' letters, but these are relatively few in number, with about six incidents over a period of seventy years. Nevertheless, they are often taken as symptomatic of a more general malaise. Given the social milieu of the eighteenth century and the circumstances at Portsmouth, where there was an abundance of liquor, money and ladies of easy virtue, and a mass of tensions between the local populace and the armed forces, a cocktail spiced by the natural exuberance of the Academy's youths, the record of incidents arguably does not suggest systemic disorder at the establishment. Again, favourable comparisons might be made with public schools of the eighteenth century, which were a byword for disaffection and violence. At Eton, for example, fights between pupils and Thames boatmen were a popular and regular pastime, and in 1768 there was a pitched battle with the butchers of Windsor, which resulted in pupils having to disguise themselves as women in order to return safely to their college.[19]

The Academy had got off to a bad start: within weeks of the first pupils arriving, the Commissioner was directed by the Admiralty

An Edwardian era view of the Royal Naval Academy: it is little changed today. (Portsmouth Royal Dockyard Historical Trust)

to enquire into the 'indecent and insolent behaviour of the young gentlemen who have been admitted'. Shortly afterwards, seven pupils aged between 14 and 16 absconded from the dockyard with a newly joined pupil named George Dashwood. On their arrival at the Sun Tavern, just outside the dockyard gate, Dashwood was forced to buy wine and brandy for the group; after two hours carousing, the group returned to the academy 'very drunk', and such was the state of Dashwood that the headmaster 'was obliged to call on the surgeon of the yard for assistance and relief'. The two older pupils held to have been ringleaders in this initiation ceremony, Peregrine Baber and Francis Colepepper, were expelled in February 1734.[20] In the following year, three pupils – Edward Rich, Richard Stacey and Edward Medley – escaped over the dockyard wall in the night, having removed their bedding, which they sold locally for £27. The purchaser informed on them, whereupon the pupils 'fell on the man and beat him up so much that he is still confined to his bed'. On the next night, four of the miscreants 'eloped who cannot be heard of'.[21]

After these early problems things seemed to settle down, with only sporadic exceptions. In September 1757, the Commissioner highlighted the practice of money lending by artificers in the yard to scholars of the Academy, which enabled them to 'pursue and establish vices, very detrimental to their own character and the peace of their friends'. Artificers were to be told they would be dismissed if they again made such loans. In June 1766, the scholars were said by the Commissioner to be spending a considerable part of their time absent without leave in public houses outside the yard, often running up large debts. Regulations were issued restricting the going out and coming in of scholars, as well as instruction to the warders of the dockyard gates.[22] James Trevenen, a scholar there from 1772–76, described bullying, idleness and debauchery in his letters home.[23] In a letter from Commissioner Gambier to yard officers in July 1774, the kitchen is spoken of as 'amazingly dirty', whilst the cook 'is not a cleanly person' and 'the scholars' heads abound with vermin'. He advised:

Care is to be taken that they appear constantly washed and combed at breakfast and dinner. Each scholar is to have a clean shirt, stock or cravat, stocking and handkerchief three times a week and clean sheets once a month. The beds are to be made by the maid servant, who is to sweep their bedrooms daily and occasionally also wash

them. Scholars are strictly forbidden to throw stones over the Dock wall. Any scholar who shall be known to go to a billiard table shall not have leave to go out of the Dockyard in future.

Gambier continued to take a close interest in the Academy. In December 1776, he informed the headmaster that, 'as a result of the late insolent and riotous proceedings' three scholars were to be expelled, whilst another seven were to be confined.[24] One dissatisfied parent, the widow of Captain Robert Faulknor, wrote:

That the Royal Academy at Portsmouth is the dirtiest school in England I know extremely well … Mr Hatt the hairdresser has sent them to London with their heads swarming with vermin. The King pays a dancing master, who perhaps comes once in six weeks and is not capable to teach them a bow.

Nevertheless, Mrs Faulknor sent three sons there.[25]

Corporal punishment was part of the regime. In March 1759, two scholars were to be whipped by the headmaster for staying out late. Mr Russell, 'the largest boy', refused to submit, saying he would rather be expelled, and the headmaster summoned the assistance of the foreman of the labourers with about a dozen men 'to keep the rest quiet', having on a previous occasion 'been hindered by a general assembly of the scholars from correcting Mr Edwards'. Mr Russell heard of their coming and escaped by jumping out of a window. The other boy, Mr Mure, got on the other side of a table and refused to submit, and other masters were called to assist. The other scholars got their knives ready to use 'in the defence of anyone who was meddled with', and one boy, Mr Cornelius (who had been expelled five months earlier but later readmitted), threatened to stab anyone who assisted with whipping. The Admiralty Board ordered Mr Russell and Mr Cornelius to be expelled, but in the following July Mr Russell was readmitted to the Academy.[26] In February 1777, Commissioner Gambier advised the headmaster, 'As a result of an Admiralty order you are to cease chastening the scholars with a rattan or stick', it being thought inappropriate for 'any young gentleman under education for a military profession'.[27]

In 1748 a sad calamity befell one of the scholars who, with another boy, had ventured out to skate on the ice-covered fishpond behind the officers' houses; the ice was not thick enough to bear his weight and gave way. Mr Roddam fell in and was drowned, 'there being no person near enough to, or capable of assisting him'. The Commissioner wrote to the Secretary of the Admiralty, saying, 'I am sorry for this unhappy misfortune, and the more so as he was a very promising youth.'[28] A distinguished Academy graduate was Philip Broke, who had attended in 1791. He achieved particular fame as Captain of HMS *Shannon* in its victory over the USS *Chesapeake* in the War of 1812. Two of Jane Austen's brothers, Francis and Charles, attended the Academy in 1786 and 1791 respectively, and both went on to become admirals.

The prejudices against the Academy graduates continued, with some captains refusing to take them. Byam Martin (who attended the Academy in 1785), whilst conceding that the teachers were excellent, thought that the Academy was 'not conducted well', that, 'there was a screw loose somewhere and the machinery did not work well' and that the best place for an officer's education was a 'well regulated warship'. The Duke of Clarence, later William IV, when a naval captain, remarked, 'there was no place superior to the quarterdeck … for the education of a gentleman.' This despite the fact that the Duke's principal educational experience aboard a warship was that his schoolmaster had twice tried to kill him. Lord St Vincent is said to have asked the parent of a prospective pupil whether he was 'so partial to that seminary as to hazard a son there?', and to have added in February 1801 that he considered the Academy 'a sink of vice and abomination [that] should be abolished'. Four years later, a successor as First Lord, Barham, was of a similar mind, describing the Academy as, 'a nursery of vice and immorality'.[29]

The Academy was not abolished but was reconstituted in 1806 and renamed the Royal Naval College. The building was extended, and a cross wing added, in the early years of the nineteenth century,[30] and the number of places raised to seventy, of which forty were reserved for the sons of officers. The College still only catered for a minority of officer aspirants: only 2.7 per cent of the officer entry between 1806–14 who survived until 1854 joined the Navy from the College, though this rose to 11 per cent after 1815.[31] In 1816, it merged with the School of Naval Architecture (which had been established in 1811 at Portsmouth: this school of twenty-four students lasted until 1834). In 1822, the College replaced Trinity House as the place where naval schoolmasters

Midshipman

A midshipman in the Georgian era. Once they had completed their studies at the Academy, the scholars had to serve two years at sea before they might be rated as midshipmen.

were examined, and in 1836, the name of this type of officer was changed to Naval Instructor.[32]

On 30 March 1837, the Royal Naval College in its old form ended and the title of College Volunteer (as the pupils were by then known) was abolished, to be replaced by that of Cadet in 1843.[33] From that date, all boys setting out on a naval officer career proceeded directly to sea. The closure of the College had created a gap in cadet training that was not filled until 1857, when provision was made on the hulk *Illustrious* moored in Haslar Creek, and then in 1859 in the hulk *Britannia*, which moved to Portland in 1862. In 1863, she was again moved, to the River Dart, and continued performing that role at Dartmouth,[34] where the shore-side Britannia Royal Naval College was opened in 1905. The Academy building at Portsmouth was turned over to the scientific and professional training of naval officers until that transferred to Greenwich in 1872,[35] and from 1906–41 was the Royal Naval School of Navigation (HMS *Dryad*). It then became a staff officers' mess, but this closed on 14 June 2007.

'a black vicious ugly customer as ever I saw. Whale-like in size, and with as terrible a row of incisor teeth as ever closed on a French frigate.'

Charles Dickens describing HMS *Warrior*, 1861

The Victorian era brought enormous expansion and change to the dockyard, taking in a huge expanse of land to the north-east, much of it reclaimed from mudflats, to create Nos 2 and 3 Basins and the surrounding dry docks and workshops. Sail gave way to steam, and wood to iron, requiring extensive new facilities, technologies and crafts. In 1845, there were still 276 sailing warships in the Navy, including eighty-seven ships-of-the-line, augmented by thirty-seven steam warships (none of which were ships-of-the-line). However, by 1860, sixty-four of the ninety ships-of-the-line were steam-assisted, and steam predominated in the fleet of 323 frigates, sloops and smaller vessels. In this era, over half of the fleet was in Ordinary, i.e. in reserve, including most of the sailing ships.[1] Unlike previous developments to the Dockyard, which had been prompted largely by wars, it was now technology that was driving change. The Navy was also growing in size to police the new worldwide territories of the British Empire, and the naval ambitions of France remained a factor. By 1843, the French were already establishing steam yards at their dockyards, the one under construction at Cherbourg being especially impressive.

Engineering Management

The Royal Navy was at first heavily reliant on private contractors for meeting the engineering needs of the steam Navy, with only limited facilities available in the dockyards, until provision was made at Woolwich Dockyard in 1838–39 with a steam factory, smithery and foundry, but even these soon became inadequate for the Navy's needs.[2] The economies of 1831–32 had led to provision at Portsmouth being run on a shoestring following the abolition of the posts of Masters of Metal Mills and of Millwrights, though the Master of the Woodmills had won a reprieve. The establishments of these departments were all reduced. George Hackney, the Superintendent of Millwrights and Engine-makers, was given the additional superintendency of the Boiler House: he was succeeded by his assistant, Robert Taplin, who by 1838 was carrying a heavy load of responsibility. He supervised all yard machinery, including the 50hp engine at the Metal Mills, the machinery for copper and iron milling, the two 30hp engines at the Wood Mills, the blockmaking machinery, turnery and sawmills, the pumps, the 6hp ropery engine with tarring machinery and cordage machinery, and a 6hp engine at the Millwrights' Shop, with lathes and boring mills. As well as all this, he was often called on to repair the machinery of steam vessels. He was effectively in Goodrich's old position of Engineer and Machinist, albeit on a much lower salary and with fewer assistants.[3] In October 1844, the Admiralty advised Taplin that apprentices to millwrights and engineers should be paid the same as apprentices to shipwrights, evidence that the importance of the new trades was recognised.[4] These apprenticeships helped the dockyard become more self-sufficient in marine engineering recruitment, since it had hitherto relied heavily on skilled men being recruited from the Thames Estuary, where private marine engineering firms were concentrated.

The yard was building steam paddle vessels with wooden hulls, but their engines were manufactured and installed by contractors. The first screw-propelled ship to be built at Portsmouth was the *Rifleman* of 1846, with 100hp engines from Messrs Miller and Ravenhill, who also fitted the screw propellers. Designed by John Fincham, she was built of wood, though for comparison purposes an iron-hulled sister ship, the *Sharpshooter*, was built by Ditchburn and Mare at Blackwall.[5] These ships still relied on sail as their main form of propulsion, with engines for additional power. Fincham was master shipwright at Portsmouth from 1844–52, and had previously been superintendent of the School of Naval Architecture there.

The Steam Yard

Following the abolition of the Navy Board in 1832, the post of Surveyor of Buildings, a civil engineer and architect who had designed dockyards since 1812, was superseded in 1847 by the use of Royal Engineers in this function, a Department of Architecture and Civil Engineering being created, later called the Admiralty Works Department.[6] In April 1842, it was decided to introduce steam machinery into the smitheries of all the dockyards.[7] Plans for a new steam yard at Portsmouth – a complex of drydocks, factory and 7-acre basin – were drawn up by a group of Royal Engineers officers led by Captain Sir William Denison and Captain Henry James, assisted by Andrew Murray, who became the Chief Engineer employed by the Admiralty at Portsmouth Dockyard. Murray was thirteen years younger than Robert Taplin, but had wider professional experience in the private sector: he was appointed Acting Chief Engineer and Inspector of Machinery at Portsmouth by a warrant dated 9 May 1846, and had principal officer status.[8]

The first stone for the new basin was laid on 29 May 1843 by Rear Admiral Hyde Parker, and five years later, on 25 May 1848, Queen Victoria was present for the opening of the new Steam Basin (now No. 2 Basin) to the north of the old dockyard area. *The Times*[9] reported on what it said was a great gala day at Portsmouth, in sweltering hot weather. A great number of the population of Portsmouth and its neighbourhood flocked to watch the event from booths and platforms which had been arranged all around the margins of the basin. Three immense standards waved from the roof of the incomplete steam factory, which was filled in many parts with spectators 'more curious than cautious'. At the entrance to the basin were the ten-gun brig *Rolla*, manned by young apprentices, the gunnery training ship *Excellent* and a host of ships and Admiralty barges. The basin edge was lined with troops from the Marine Artillery, and portions of the 9th and 17th Regiments, numbering 1,050, as well as 2,078 men of the Royal Dockyard Brigade. The latter force consisted of dockyard workmen, who had undergone a course of drilling, and were supplied with

The impressive Steam Factory, facing the Steam Basin, seen c.1975, was completed in 1849 to house numerous machine tools to meet the needs of the nascent steam navy. (Portsmouth Royal Dockyard Historical Trust)

regimental uniforms. In a rather veiled compliment, *The Times* said, 'The appearance they made was very creditable, and we were particularly struck with their resemblance to continental troops in the absence of that stiff and perfect precision which is the peculiar characteristic of the British soldier.' Six or seven bands were present, but they 'showed no particular anxiety about playing in tune or out of tune, or even about playing at the same time entirely different tunes'. The Mayor and the Corporation, in their scarlet gowns lined with fur, looked uncomfortable in the heat, and on the uncovered platforms spectators took refuge under a sea of brown parasols. At about 3.30 p.m., the new iron screw yacht *Fairy* steamed into the basin, carrying the queen and her consort, followed by a procession of beautifully painted and manned barges, and the bands struck up in succession 'God Save the Queen' as the yacht made a circuit of the basin before berthing a little above the north inlet, where the principal booth was erected. There the Duke of Wellington and a host of other worthies joined the ceremony as the monarch witnessed the laying of the last stone. Afterwards, the royal party re-embarked at 4.22 p.m. in the *Fairy* for Osborne House on the Isle of Wight. At 5 p.m., the Dockyard Brigade sat down to an 'excellent cold dinner' provided by Mr Wood, army outfitter, of the Common Hard. At the same time, the contractor's workforce, numbering 1,600, sat down to a similar banquet provided by Mr Bleaden, of the Poultry, London, the bands playing them in to the tune of *The Roast Beef of Old England*.

Excavation of this non-tidal basin had employed 1,200 convicts from hulks moored in the harbour, as well as 1,050 wage labourers. The entrance to the basin was closed by a wrought iron caisson, but this proved inadequate and was replaced by another installed to the east and coming into service in 1876. One of the cranes built to serve the Steam Basin was the first steam-powered crane in the Dockyard, and had a curved metal jib (which became known as a swan neck), whilst the two others were still manually operated.[10] Two months before the Steam Basin opened, the last sailing ship to be built at Portsmouth was launched. She was the *Leander*, a fifty-gun Fourth Rate designed by Robert Blake, Master Shipwright at Portsmouth from 1835–44.[11] Three new dry docks eventually opened into the Steam Basin: Nos 7 and 8 were completed in 1849 and 1853 respectively, in accordance with the steam yard plans (though No. 7 Dock, also known as the Camber Dock, was a single

dock rather than the originally planned double dock); No. 9 Dock, completed in 1850, opened into the harbour at the north-west corner of the yard (it was closed c.1898 and filled in during the Edwardian period to make room for a steel plate stacking area for No. 5 Slip). No. 10 Dock was added to adjoin No. 7 Dock on its west side, opening into the harbour, and completed in 1858.[12] The size of the new docks reflected the increase in size of warships – No. 7 was 28ft longer than Bentham's last dock (No. 3) to accommodate the longer wooden ships built with the use of Seppings diagonal framing, and had an entrance that was 23ft wider than that dock, because of the advent of paddle-wheeled vessels. No. 8 Dock (or South Inlet) was 42ft longer than No. 7 but had a narrower entrance, being designed for screw-driven vessels. Its entrance had straight sides on a slight incline rather than the elliptical arched entrances of Nos 3 and 7 Docks. All subsequent docks had similar entrances, which allowed the caisson to be more speedily placed in position and was more suitable for the shape of paddlers.[13] The double dock arrangement created by Nos 7 and 10 Docks allowed even longer ships to be docked, including HMS *Warrior*, the radical iron-hulled frigate which arrived in Portsmouth for the first time on 20 September 1861, and it was in those docks that she was refitted and rearmed between 1864 and 1867.[14] This was the only facility at Portsmouth that could dock the new generation of iron-hulled ships, pointing to the need for even larger docks. The third dock to open into the Steam Basin, No. 11 or North Inlet, was completed in 1865 and was 67ft longer than No. 8, the South Inlet. At its head was a 90hp pumping engine house.[15] Although the largest single dock built to date, this was only a stopgap before much larger provision could be made in the Great Extension. In 1858, Nos 2 and 4 Docks were each lengthened at their heads to 263ft 6in, and a caisson was substituted for the gate in each dock.[16]

The Steam Factory, built along the west side of the Steam Basin, was a huge engineering workshop in which a wide range of machining and fabrication operations could be performed. Completed in 1849, to a design by Captain Henry James RE, it transformed the Dockyard's capabilities, harnessing the tools of the Industrial Revolution. Its appearance is impressive, the design being an architectural riposte to the storehouses being built by the French Navy at Cherbourg. At 687ft long (a length only exceeded by the ropery) and 39ft wide, with two storeys, the upper one

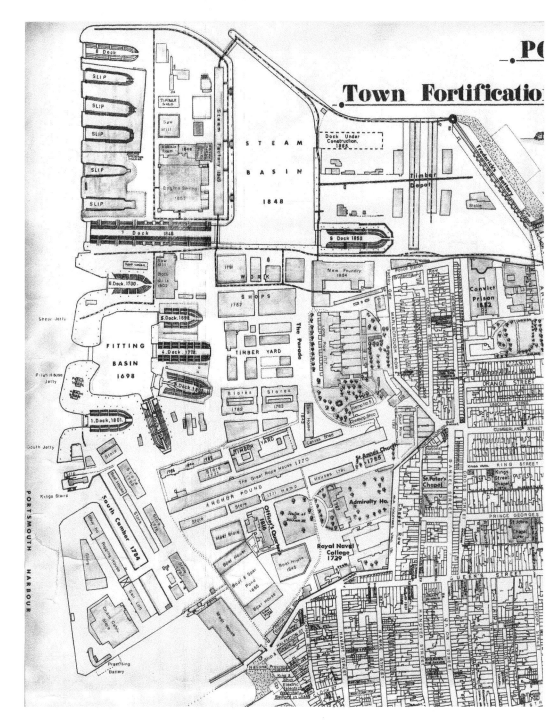

Detail from a map showing a plan of the dockyard c.1865, with the new Steam Basin and its associated buildings, at the north end of the old yard. No. 11 Dock is still under construction at the north-west end of the Steam Basin. Nos 7 and 10 Docks are shown as one double dock (labelled No. 7), a configuration that enabled HMS *Warrior* to be docked for a refit between 1864 and 1867. The convict prison, which supplied labour to the Navy Works department, is shown on the right of the plan. (Portsmouth Royal Dockyard Historical Trust)

The Smithery was completed in 1855 to the west of the Steam Factory.
(Portsmouth Royal Dockyard Historical Trust)

The Iron Foundry dated from 1861 and was built in a similar architectural style to the Steam Factory.

being carried on massive cast-iron beams, it is built of deep red brick offset by Portland stone cornices and pedimented bays. On the ground floor (from south to north) were the heavy turning shop, the erecting shop, the punching and shearing shop and the boiler shop. On the upper floor were the millwrights' shop, the light turning and fitting shop, the pattern makers' shop and the pattern store.[17] An 80hp Engine House was built to the west of the Steam Factory it served (but has long vanished). In 1862–63, the Steam Factory workforce numbered 750.

The Iron and Brass Foundry (1848) facing No. 8 Dock was designed by James and built in a similar style to the Steam Factory, though in 1861 it lost its iron founding function with the completion of the adjacent Iron Foundry. A new Smithery was erected behind the Steam Factory, using a more economical metal frame clad with iron sheeting, and within it was a huge Nasmyth steam hammer. This was designed by James and built under the supervision of his successor, Colonel Godfrey T. Greene, who became Director of Admiralty Works in 1850. Work began in 1852 and was completed by September 1855.[18] A new sawmill was also built in 1848, taking over much of the work of Bentham's woodmills. The new Iron Foundry and Pattern Makers' Shop (1861) was designed by Greene in a similar style to the Steam Factory (it was given a sympathetic extension in 1873 when the stables belonging to the adjacent officers' terrace were demolished to make space). A handsome two-storey building of red brick with stone detailing and iron-framed windows, it has four wide main entrances allowing access for the new dockyard railway, with arches of Portland stone strengthened by iron transoms.[19] The Armour Plate Workshop, later No. 1 Ship Shop, was completed in 1867 on the east side of the Steam Basin. Its architectural style was similar to the Steam Factory and the Iron and Brass Foundry, but was enhanced by Italianate towers pierced by slim windows and surmounted by a finial. It was enlarged in the 1880s,[20] and part of the building has survived, incorporating the former Light Plate Shop (1880), which was later known as No. 1 Ship Building Shop.[21]

Although most of the new facilities were built around the Steam Basin in the 1840s, the elegant new Boat Storehouse, now No. 6 Boathouse, was constructed in the older part of the dockyard, to the east of the Mast Pond, and completed in 1844 to a design by Lieutenant Roger Beatson of the Royal Engineers, who was appointed superintendent for the Admiralty Works Department at Portsmouth at the age of 27 in 1839. Beneath the traditional buff-coloured brick exterior of the boathouse is a massively over-engineered three-storey frame of iron columns and girders. The front elevation has three large pairs of arched doors, through which boats could be hauled on rails from the Boat Pond, and large rectangular windows, many of which were originally fitted with wall cranes. Inside, a central well allowed boats to be hoisted to the upper floors.[22] At this time the mast pond was flanked by two mast-making houses on wooden piles. These were rebuilt on iron piles in 1875 and 1882, becoming Nos 7 and 5 Boathouses respectively. At the same time as Beatson's new Boat Storehouse, a Chain Store was built in a single-storey iron-framed building to the design of Denison and Beatson to the west of the Sail Loft. This was a new requirement in the dockyard, accompanying the widespread use of iron chain cables. From 1909 until the 1970s, the building served as the Chain and Cable Test House and Store.[23]

Completed in 1843, the Water Tank was supported by hollow cast-iron columns which also served as water pipes. The enclosed area underneath it later became a fire station.

Beatson also designed a new iron Water Tank in 1842: sited between the Ropery and Long Row, and completed in 1843, it had hollow cast iron columns as water pipes to convey water from the raised cistern into the ring main. The open sides underneath the tank were closed with corrugated iron sheeting in 1847 to form a wood store for deal, but the building later became a fire station, losing its water tank in 1950.[24] A large new storehouse, No. 12 Store, was built around 1855 to the west of the Middle Store (No. 10), with iron beams to support the upper floors. Of red brick, it boasted a Portland stone string course between the round-headed windows and the decorative coving and parapets.[25]

The Dockyard Railway

Wooden tracks had apparently been used in the dockyard since the sixteenth century to haul masts in trucks pulled by horses from the Mast Pond to masthouses, and then to the sheerlegs, to be stepped in the warships.[26] A tramway system was in use by 1828, when convicts from the hulks were ordered to take over the haulage of trams from horses. The trams ran on parallel slabs of flat granite some 2ft 3in wide, separated by Guernsey pitchers 2ft 10½in across set in a concrete bed. A standard gauge railway, probably worked by horses and capstans, was laid down around the new Steam Basin, and the granite tramway was extended to meet it at a turntable south of the caisson to the Camber Dock. A branch of the tramway ran east to the timber stacking ground and the casemates of Frederick's battery.[27] The dockyard railway was linked to the main line system in March 1857 by a line which ran from Portsmouth Town Station, skirting Victoria Park and crossing Edinburgh Road, to enter the yard near the Unicorn Gate.[28] However, steam engines could not pass beyond the reception yard (because of numerous turntables installed at the right-angled bends on the roadways), and movement of railway trucks inside the dockyard was by horses until 1869, when the first steam locomotive was acquired to take up some of the duties. In the 1870s, the granite tramway was still in use, running to Nos 1–3 Docks, but eventually the railway system extended to most parts of the yard and most of the turntables were removed. This allowed the Admiral Superintendent and distinguished visitors to travel in style in a four-wheeled carriage known as the Special Saloon.[29] A second line into the dockyard opened in January 1878 on a viaduct from Portsmouth Harbour station across the Common Hard to Watering Island Jetty (now the South Railway Jetty), with a swing bridge to give boats access to the Hard. This was a prestige line for occasions of state or embarking troops. Queen Victoria used the link when travelling to Osborne House, and an elegant cast iron royal railway shelter was built in the dockyard and erected on the South Railway Jetty in 1893, although two decorative railway waiting rooms had been erected when the line was installed.[30]

Shipbuilding

Shipbuilding capacity in the yard was expanded through a series of developments between 1838 and 1850. A new slip, No. 4, was built in 1838, followed by another about ten years later. The largest ship-of-the-line built by then at Portsmouth (and the last three-decker to be built for the Navy) was the 110-gun *Queen* of 1839. She had been laid down as the *Royal Frederick*, but was renamed before her launch in honour of the new queen. In 1842, Queen Victoria went aboard the *Queen* at Spithead, during her first visit to Portsmouth after accession. She had arrived at Portsmouth by coach and four, and stayed at Admiralty House. Accompanying her were Prince Albert and the aged Duke of Wellington.

Wooden roof covers on building slips were very prone to rot and were also a fire hazard. In 1845–46, the roof covers on Nos 3 and 4 Slips were replaced by fireproof iron roofs, which combined cast and wrought iron frames covered with corrugated galvanized iron (CGI) sheeting with skylights, the work being undertaken by George Baker & Son of Lambeth. CGI had been patented on 30 June 1829 by Henry Robinson Palmer, architect and engineer to the London Dock Company. A similar treatment was given to the new No. 5 Slip, which also had a raised roof section at the landward end to facilitate work on the ships' bows. The new slips led to an increase in the building programme at Portsmouth, with seven First

The First Rate *Marlborough* was launched at Portsmouth in 1855 and is seen here in Grand Harbour, Valetta, Malta, as flagship of the Mediterranean Fleet.

and Second Rates being launched between 1855 and 1860. No. 5 Slip was to be extended to a length of 666ft in the Edwardian era, by which time the other slips were no longer used for shipbuilding and No. 4 Slip had been filled in to become a Shipbuilding Shop.[31]

Whilst paddle propulsion had been thought unsuitable for ships-of-the-line, because of the paddles' vulnerability to damage in action and the loss of part of the broadside space, the invention of the screw propeller meant that these ships could become steam-powered. Under a programme announced in December 1851 to provide the Navy with a steam-driven battle fleet, the design of four First Rate sailing ships on the stocks was modified by the new Surveyor, Captain Baldwin Walker. At Pembroke, the *Windsor Castle* was cut apart in two places on the stocks in January 1852, lengthened by 30ft overall and given screw propulsion with 780hp engines designed and built by Robert Napier. On 31 July 1855 at Portsmouth, the 131-gun First Rate *Marlborough* was named by Queen Victoria in a launch ceremony, but became stuck on the ways and could not be dislodged until midnight, and then with some damage.[32] The *Marlborough* was begun and framed as a 110-gun ship-of-the-line, but was converted into a steam-propelled screw ship by lengthening her whilst still on the stocks, following the example of the *Windsor Castle* (which became the *Duke of Wellington* in honour of the man who had died on the day of her launch).[33]

On 25 April 1857, a third ship, the *Royal Sovereign*, was launched at Portsmouth after six and a half years on the slip, during which time she too had been cut in two, and lengthened by 23ft amidships and 7ft at the stern for the screw aperture, to also become a screw-propelled 131-gun steam line-of-battleship,[34] and was placed in Ordinary. Like *Marlborough*, she had been ordered as a 110-gun sailing ship-of-the-line, but whilst on the slipway she was ordered on 29 June 1848 to be lengthened by 6ft, to carry 120 guns. On 23 June 1854, another order required her to be lengthened by a further 30ft and converted into a screw ship-of-the-line of 131 guns. On completion she ran trials in Stokes Bay and achieved a speed of 12.25 knots unrigged.[35] On 4 April 1862, to try the idea of Captain Cowper Coles, the Admiralty ordered a much more radical conversion of her at Portsmouth into the Navy's first turret ironclad, the only one ever to have a wooden hull. She was cut down to the lower deck, giving her a freeboard of only 7ft, her 3ft thick oak hull was sheathed with 5½in iron plate and she was re-armed with five

10.5in 300-pounder muzzle-loading guns in three single and one double centre-line turrets, on the top of the hull, rather than the broadside arrangement of guns mounted internally on carriages which the *Warrior* still had. *Royal Sovereign* was the first capital ship in the Navy to dispense with sails for propulsion, having three stumpy pole-masts with jib sails to steady her roll at sea when the engines were stopped. Her two-cylinder 2,460hp Maudslay engines gave her a speed of 11 knots in Stokes Bay on 23 June 1864.[36] She was commissioned at Portsmouth on 7 July 1864, but the experiment was not repeated: she was the only one of the four ships to be cut down as a turret ship, it being more economical to build new iron-hulled turret vessels. She had a limited range, carrying only 300 tons of coal,[37] and her low freeboard made her unsuitable for deep-sea cruising.[38] She did not fit in with the main fleet and her short service of nine years was spent on gunnery trials and training.[39] However, the *Royal Sovereign* did demonstrate the practicability of gun turrets and their advantages over the broadside arrangement.[40] The fourth ship, *Prince of Wales*, was also built at Portsmouth and lengthened on the slip. Ordered on 29 June 1848, she had commenced her conversion to steam on 27 October 1856 and was launched on 25 January 1860. On trials in Stokes Bay, she made 12.6 knots unrigged.[41]

The First Rate *Frederick William* was initially ordered from Portsmouth Dockyard on 12 September 1833 as a 110-gun Queen-class ship-of-the-line, under the name *Royal Sovereign*. The order was suspended on 7 May 1834, but was later renewed, and she was renamed *Royal Frederick* on 12 April 1839. She was laid down on 1 July 1841, but work proceeded slowly, and on 29 June 1848 she was reordered to a modification of the Queen-class design, still powered by sails alone. The design was again modified on 28 February 1857, when it was ordered that she be completed as an eighty-six-gun screw battleship. Conversion work began on 28 May 1859, and the ship was renamed *Frederick William* on 28 January 1860 (on the occasion of the Princess Royal's wedding to Prince Frederick William, who later became the German emperor) and was launched on 24 March 1860. She was completed in June 1860.[42]

Another First Rate, the 120-gun *Victoria*, was the last British wooden three-decked ship-of-the-line commissioned for sea service and the first to be started in construction as a steam ship. With a displacement of 6,959 tons, she was the largest wooden battleship which ever entered service and was also the world's largest warship until the completion of *Warrior* in 1861. *Victoria* was ordered on 6 January 1855, laid down on 1 April 1856 at Portsmouth, and launched by Princess Frederick William in the presence of the eponymous queen at a glittering occasion on Saturday, 12 November 1859.[43] During trials in Stokes Bay on 5 July 1860, she reached a top speed of 13.14 knots. It was remarked that her engines worked well, with an abundance of steam.[44] Her Maudslay engine was powered by eight boilers and developed 4,293hp.

Haslar Gunboat Yard

Haslar Gunboat Yard was an annex to the dockyard, built on the Gosport side of the harbour, that was opened in 1857 for the storage in reserve, repair and building of steam gunboats after the Crimean War (1854–56). During that war, gunboats were hastily built (mostly by private contractors) in large numbers for deployment in the Baltic and Black Sea. Often built with poor workmanship and employing 'green' unseasoned timber, they were frequently liable to extensive dry rot. To relieve the dockyards of the task of maintaining them, sites at Plymouth, Chatham and Haslar were considered for a gunboat yard before Haslar was chosen. Built on former farmland facing Haslar Creek above Haslar bridge, the yard had covered boat sheds in a continuous line, served by two slips and a steam locomotive traverser, known to staff as the Elephant, to move the boats on cradles laterally from the slipways to the sheds. The design of the traverser was in 1856 attributed to Colonel G.T. Greene, the then Director of Engineering and Architectural Works, and his deputy, William Scamp, but in an 1857 publication the designers were said to be Scamp and Thomas White, shipbuilder of Portsmouth. A traverser system was already in use at the locomotive sheds in Swindon, and was designed by Daniel Gooch, who had been appointed by Isambard Kingdom Brunel. (This association may have led to the Haslar traverser's design being wrongly attributed to Brunel in some sources.) The slip at the Haslar yard was 130ft long, laid with tracks, upon which boats were hauled up by a steam winch to the traverser line with fourteen parallel rails, 1,400ft long, served by the Elephant locomotive, above which were the berthing sheds, each 115ft by 30ft, into which the boats were hauled. They were covered by corrugated iron roofs supported by iron frames and columns 18ft high. Every ten of the berths were separated from the remainder by a brick firewall, indicating that his was a highly valued site.[45]

The first attempt at hauling up a vessel, the *Gnat*, on 25 November 1856, was not entirely successful as the machinery was not powerful enough to haul the vessel all the way up. Modifications were made and the first successful attempt was completed in January 1857. Additional sheds were being built so that by 1859 there were forty of them, with berths for forty-seven boats. Some additional, smaller sheds were built of timber for torpedo boats, and there were also repair workshops, storehouses, a mast house, a house for the master shipwright, a guardhouse and a small police barracks, and the complex was surrounded by a high brick wall with guard posts at each corner and a parapet for patrolling. The yard measured 2,640ft in length and 816ft in width, the length of the shed line being 1,900ft.[46]

In 1860, it was reported that forty-seven gunboats were in the sheds, of which twenty-two had been repaired and with the exception of coppering were ready for launching, nine were under repair, fourteen were awaiting examination and repairs, five of which were uncoppered, and two had only just been hauled up.

Haslar Gunboat Yard in 1878, showing gunboats in the sheds on the left and the steam traverse on the right.

Of the boats that had been repaired or examined, only two had had sound hulls (the *Earnest* and *Escort*, both built by Patterson of Bristol); many of the others were in a deplorable state.[47] By 1863, the situation had changed little: *The Times* reported:

> This precious flotilla ... this squadron on which so much money had been spent, was found rotten and good for nothing, and the boats were only fit to break up or crumble to the dust where they lay. They were not broken up, and were delayed at an immense expense in falling literally to pieces. Two certainly, the *Caroline* and the *Mackerel*, were eventually condemned and taken to pieces, but even now fragments of their poor rotten skeletons may be met with in obscure corners of the Haslar Yard, each fragment forming part of a tale of fraudulent copper bolt fastenings and sappy wood.

Seventeen of the forty-seven boats then in the yard had been condemned for breaking up, whilst another fifteen were to be surveyed for sale or breaking up, nine were to be repaired and two were to be converted into coal depot vessels to convey fuel to ships at Spithead (the remaining four were not mentioned). Another 110 gunboats were afloat in home ports or on foreign stations, 'all of which must be looked upon by the lights of our present experience as possessing a still more unfavourable average condition than even those quoted above as in the Haslar list'. The repairs made to the boats at Haslar were said to be superficial, enough to make the boats look presentable, but masking more fundamental defects. However, it seems that the use by the builders of unseasoned timber had been officially condoned by the Admiralty's Chief Constructor because the Navy's store of seasoned oak was almost exhausted in 1855 and the gunboats were thought to be expendable, unlikely to be retained in the post-war fleet.[48]

In 1860–61, six wooden gunboats were listed for construction at Haslar Gunboat Yard, these being *Bruizer*, *Cherub*, *Minstrel*, *Cromer*, *Orwell* and *Netley*, of the Britomart class. Five of them were in frame by January 1862, but their completion was delayed, probably to ensure that the wood was fully seasoned: they and the sixth vessel were all launched between 1865 and 1867. Another three were cancelled and broken up on the stocks in 1863. The Britomarts were large enough to make ocean passages for service on overseas stations and, like the earlier classes, they retained a

sailing rig for long-distance cruising.[49] By 1869, the yard was in decline, with most of the gunboats broken up, and by 1871 at least twelve sheds had reportedly been moved to Portsmouth dockyard, others had been demolished and only ten sheds remained. *The Times*, still doggedly covering the scandal of the gunboats and their yard, reported:

> Where the 40 covered sheds stood is now a chaos of uprooted timber slabs, upon which the old gunboats stood, sprinkled over with heaps of old iron and other debris, with half-starved cattle and goats in full possession, endeavouring to browse upon the scanty crop of thistles and tufts of grass which is gradually and slowly taking possession of the shingly ground.[50]

But ironically, new iron gunboats and torpedo boats were being built and some were stored at Haslar in the remaining sheds.[51] In the 1880s, ten new sheds were constructed (probably under the 1886–87 estimates) and the creek was dredged; a 1909 plan shows a total of twenty-six boat sheds.[52]

The Haslar yard continued in use repairing a variety of small craft well into the twentieth century, albeit at a reduced level of activity except during the two world wars, when it was particularly busy repairing and maintaining coastal forces craft. It finally closed in 1974, when twenty boat sheds still stood. The master shipwright's house and some of the sheds were then demolished. The ten remaining gunboat sheds are all from the original phase of construction in 1856, and other maintenance workshop sheds, ancillary buildings and the surrounding wall also survive.[53] In 2014, the site was sold for commercial and residential redevelopment.

Iron Ship Construction

In the 1860s, Portsmouth was still building wooden-hulled ships, but this began to change in 1867–68 when three composite-hulled steam gun vessels were built, with iron frames and wooden planking. In 1870, the first wholly iron-hulled ship to be built at Portsmouth, the steam gun boat *Plucky*, designed by John Fincham, was launched.[54] These ships still retained a sailing rig. It was not until 12 July 1871, eleven years after the *Warrior* was launched, that the yard launched its first large iron-hulled warship – HMS *Devastation* – and like *Warrior* she was a milestone in capital ship development. She was the first true mastless capital ship in the Navy and had four 12in guns mounted in twin turrets on top of her hull. *Devastation* was built in No. 8 Dock, and a block model of the ship showing the arrangement of plating and drawings showing the disposition of bulkhead plates and the arrangement of the draining and pumping were prepared by Pembroke dockyard officers for use by the Portsmouth officers.[55] Pembroke was more experienced in iron ship construction, but although laid down five months before *Devastation*, her sister ship *Thunderer* was launched at Pembroke eight months after her and was taken to Portsmouth for fitting out since Pembroke lacked these facilities. *Devastation* was launched by the wife of the First Lord of the Admiralty, George Goschen, and *The Times* reported:

> The ship being already afloat in the dock, and with hawsers made fast to her stern in readiness to haul her out into the waters of the steam basin, Mrs Goschen at once dashed the suspended bottle of wine against the ship's stern, and said in distinct and full voice, 'Success to the *Devastation*.' The master shipwright (Mr Robinson) then handed to Mrs Goschen a small and elaborately finished steel 'tomahawk', with which she cut the rope that was supposed to hold the *Devastation* in dock. The bands of the Royal Marine Artillery, on the fore turret of the ship, and the Royal Marine Light Infantry upon the after turret, struck up the National Anthem, the boys of Her Majesty's ships *St Vincent* and *Excellent* (present in great force) hurrahed tremendously, and the *Devastation* slowly glided out of the dock, a great addition to the present strength of the Royal Navy.[56]

The advent of iron shipbuilding in dockyards required changes to the workforce structure. In private shipyards it was the platers and their helpers who dominated iron shipbuilding, partly because the shipwrights disliked working in metal and preferred to work in wood. The platers were assisted by unskilled men who were trained to be riveters, holders-up, etc, and the whole group was organised by the boilermakers' union. When iron shipbuilding started in the dockyards, the Admiralty naturally hired platers, but in 1863 they struck (at Chatham, where the first dockyard-built iron ship, *Achilles*, was being constructed) for higher wages as part of a national movement by their now powerful and militant union,

The turret ship HMS *Devastation* was the first large iron-hulled warship to be built at Portsmouth, and had two twin 12in gun turrets on the upper weather deck. She was the first 'mastless' capital ship (i.e. without a sailing rig) and her low freeboard made her look somewhat ungainly.

and were dismissed by the Admiralty.[57] The dockyard shipwrights seized the opportunity to avoid their role becoming redundant, and extended their work to include working in both wood and metal in ship construction. However, their wages (4s 6d per day) were much lower than wages (7–8s per day) of the platers they had replaced, despite promises that they would be rewarded for retraining in the new skills: this remained a source of grievance for many years.[58] New skilled labourers were introduced in 1876 to cover the less complex skills of iron shipbuilding in functions which included drillers, riveters, holders-up, painter's assistants, iron-caulkers, boilermaker's helpers, stokers, engine drivers and wiremen, and were usually paid on piecework. A small number of ordinary labourers were still employed in fetching and carrying. This structure contrasted with that in private yards, where iron ship construction remained the preserve of boilermakers, and jobs such as riveters had trade status.[59]

In dockyards the boilermakers were confined to their nominal craft, and other new trades such as fitters, pattern-makers, founders and millwrights were also involved in metal-working. The rates of pay in dockyards remained lower than for comparable jobs in the private shipyards, partly because of the job security given by the established, pensionable status of many dockyard employees (which was still not mirrored in private yards) and the relative job security even of hired men, both being largely immune from the trade cycles which affected employment in the private yards, and partly because the Admiralty did not allow private shipbuilding to spring up in the neighbourhood of dockyards and could therefore remain in control of wage rates. In 1891, for example, the weekly wages of shipwrights employed on government work in private yards ranged from 34s 1d (on the Clyde) to 42s (on the Thames), whilst dockyard shipwrights earned between 30–34s per week.[60] Following a call by the Trades Union Congress in 1892 for the length of the working day to be reduced to eight hours and for union wages to be paid, and a resolution in Parliament that the government should set an example to private industry, the eight-hour day was implemented in the dockyards in 1894, and unskilled labourers, who were poorly paid, were given a pay rise.[61]

The Great Extension

By the early 1860s, following the introduction of the *Warrior* and her sister ships, it was clear that the Steam Basin and its surrounding facilities were becoming inadequate: *Warrior* could only be docked in Nos 7 and 10 Docks, used in tandem. Following outline approval of the project by Parliament in 1864, a scheme for a massive extension to the dockyard on its north-east side was drawn up by Colonel Sir Andrew Clarke, RE, Greene's successor as Director of Admiralty Works. It involved the construction of a tidal basin leading to two large locks, which gave entry to three new interlinked non-tidal basins and three dry docks. Work on the Great Extension, as it was called, began on 30 April 1867, increasing the total size of the yard from 116 to 300 acres. The 9-acre Tidal Basin gave access, through the new South and North Locks (later known as A and B Locks respectively), at any state of the tide, to the 22-acre Repairing Basin. It was intended that ships would proceed through the sequence of non-tidal basins as work on them progressed, from the Repairing Basin into the 15-acre Rigging Basin and thence to the 15-acre Fitting Out Basin. The caissons between the non-tidal basins were operated on the sliding principle, as were those at the entrances to the locks on the tidal side; however, the caissons between the locks and the Repairing Basin were conventional floating structures. Two of the three new dry docks (Nos 12 and 13) were entered from the Repairing Basin, whilst the other, Deep Dock (later No. 9 Dock), was accessed directly from the Tidal Basin. The three docks were of broadly the same dimensions, with lengths between 410 and 417ft, but the Deep Dock had an entrance that was 41ft 6in deep at high water springs, with a sliding caisson, whereas Nos 12 and 13 were 33ft 4in and 33ft 10in deep respectively, with conventional caissons. The two latter docks were lengthened in 1903 and 1905 respectively.[62]

The monumental and ornate new Pumping House, which was completed in 1878,[63] had two large steam engines, with a total capacity of 1,000hp, for the pumps, and two pairs of 120hp steam engines which drove air-compressors, making compressed air the main source of power in the new extension, powering the dockyard fire main and assisting the bigger engines with dock drainage. The air compressing machinery took up slightly more than half the area of the engine house, with pumping machinery in the other part. A low boiler house, with a tall central chimney, was built on the side of the engine house. The air pipes extended round the three basins, with a total length of about 2.7 miles, varied from 3–12in diameter and were connected to forty 7-ton capstans, five 20-ton swan neck cranes and to the machinery for working seven caissons and numerous penstocks.[64] In addition to the cranes, sheer legs were provided for raising heavy loads.

So ambitious was the scope of the Great Extension project that no further extension would prove necessary, and the basic configuration of the dockyard has remained unchanged right up to the present day, albeit with modifications made during the Dreadnought era. The scheme was largely undertaken by the contractor Leather and Smith & Co., headed by John Towlerton Leather (1804–85), who had been born into the third generation of a family of engineers and had become one of the leading civil engineers and entrepreneurs of the period. His biggest projects were not for the faint-hearted – the Portland breakwater, the Spithead

The massive No. 1 Pumping Station was completed in 1878 to serve the dry docks of the Great Extension.

number of chain pumps. The need for such a deep well was to ensure the workings of the lowest foundations would be kept dry. Between the two dams, which consisted of simple sheet-piles, the harbour wall was constructed.[65]

The whole area to be reclaimed was composed of soft mud, inaccessible to all means of transport. To excavate the deep mud, a series of wooden viaducts were built, on which the steam dredging machines, steam grabs, pile drivers and steam locomotives operated. There were 9 miles of railway tracks and roads serving the pile drivers and the huge Goliath steam bucket dredgers and steam shovels which discharged mud in railway wagons for transit across Fountain Lake, on one of the viaducts, to be dumped on the exposed mud bank which created Whale Island, home of the Navy's gunnery school. This viaduct had a 50ft iron swing bridge at its centre to allow shipping into Flathouse and Rudmore quays. There were also 4 miles of track for self-propelled steam gantry cranes, which received granite blocks from railway wagons and lowered them into position on the dock and wharf walls. The excavations were kept drained by steam pumps, with around eighty steam engines at work at the peak. Thirteen steam locomotives, each weighing up to 16 tons, were employed for haulage on the site. The extension project employed some 1,500 workers and 700–800 convicts, the latter being housed in a specially built prison just outside the dockyard wall (it could hold 1,020 men and was built between 1850 and 1853). Clay dredged from the excavations was utilised in the large brickworks, which at one point made 20 million bricks a year. In all, 155 million bricks were used, as well as 4 million cu.ft. of Portland stone, 5.3 million cu.ft of granite, 5,000 tons of iron, 281,000 loads of timber and 650,000 tons of concrete. The extensive use of concrete was pioneering because this material was relatively new and untested in big civil engineering projects. Twenty million cubic yards of material was excavated and 240,000 tons of material was dredged. That this preliminary work was challenging was attested by Sir Andrew Clarke at a meeting of the Institution of Civil Engineers in February 1881. He recognised the bold way in which the contractors had proceeded:

It was astonishing to see so large an amount of temporary works executed before any permanent work had been attempted. This temporary work, however, secured the rapid, certain and economical

sea forts and, largest and most demanding of all his projects, the Great Extension at Portsmouth. Because of the size of the contract, and his own advancing years (he was 63 when work started and 70 when it was complete), he formed a partnership with George Smith of Pimlico. The project was supervised for the Admiralty by Colonel C. Pasley, who had succeeded Clarke as Director of Admiralty Works, and presented huge engineering challenges, particularly as about 95 acres of the new land was on soft mudflats which were subject to tidal conditions, and all the work had to be carried out from timber stages and viaducts mounted on piles. It necessitated the building of two dams to enclose the area from which tidal water could be excluded, the outer one being 4,100ft long and the inner one 3,600ft, and the pumping dry of the enclosed area. To do this, the dams were closed at low tide and a 13ft-diameter cylinder of cast iron was sunk 70ft into the ground, the lower 18ft being of wrought iron. Above the top of the cylinder, a stage was erected to support four 20hp portable steam engines, operating the same

Work on the Great Extension, looking north across the Repairing Basin and dry docks. The entrances to the four new dry docks (Nos 12–15) can be seen from centre to left.

construction of the whole of the permanent work. The contractors were men of considerable capital, and had the boldness to use it; for few men would have had the courage to expend on temporary operations which were not unattended with considerable risk and danger, close on a quarter of a million sterling, before commencing their permanent work.[66]

The docks were built on a layer of cross-braced piles, with sheet piling enclosing the work. Portland concrete was then laid to give a base for brickwork in Portland cement. Finally, the top layer, which was the bottom of the dock, was of granite. In No. 13 Dock, the thickness of the flooring from the concrete to the granite floor is 11ft 6in. The sides of the dock were built up in layers of brick and concrete, then faced with Portland stone with granite toppings to the alters and copings. All the docks and locks were completed with granite slides for the removal of material from the dock bottom when refitting ships. The flooding of the docks was direct through culverts from the Repairing Basin, or in the case of the Deep Dock, the Tidal Basin. The two locks could be flooded from either the Tidal Basin or the Repairing Basin. Drainage was through a network of culverts

to the main pump station.[67] The locks, docks, Repairing Basin and Tidal Basin were completed in 1876, whilst the other two basins followed in 1881. The Great Extension lacked major workshops until 1905, though space was left for them, and had to rely on the dockyard railways bringing materials from workshops and stores that were mostly in the older parts of the yard.[68] However, advances in armaments called for specialised buildings: the Gun Mounting Shop, built between the Power Station and No. 9 Dock, was completed in 1886, as was the Torpedo Shop, situated to the east of No. 12 Dock.[69] On a knuckle at the south-west end of the Fitting Out Basin, a Coaling Point was constructed with space for 18,000 tons of coal, and quayage for the bunkering of vessels using hydraulic cranes which could discharge at 500 tons per hour. This was around double that which could be achieved with manual labour, but only

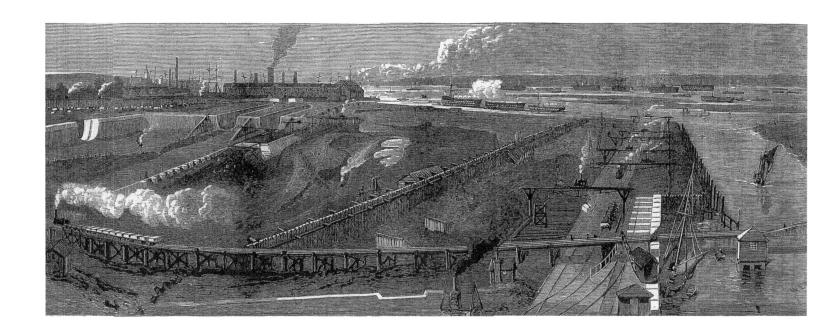

Looking west across the Rigging and Fitting Basins as work proceeds on the Great Extension. The railway viaduct in the foreground leads (out of the picture to the right) to Whale Island, where the excavated mud was deposited.

cruisers could coal alongside, and battleships still had to be coaled by lighter.[70]

The first phase of the dockyard extension was officially opened on 27 April 1876, on the same day that the battleship *Inflexible* was launched at No. 5 Slip by Princess Louise of Lorne. The work to complete the Rigging and Fitting Out Basins was still incomplete at that time. The launch ceremony was attended by the usual pomp and ritual of such events. The dockyard gates were thrown open to the public about 10 a.m., and at 11.30 a.m. the clamour of numerous bells signalled to the workmen that they could cease work for the rest of the day. The warships in dock and in the harbour, and the various yachts at their moorings, were clad with streamers from stem to stern, composing 'the prettiest sight of the day'. The spars and rigging of the men-of-war were lined with sailors, and

in the dockyard the bands of the Royal Marine Artillery and the 52nd Regiment played lively music. The royal stand at the prow of the *Inflexible* was carpeted with crimson cloth and accommodated both the royal party and the Lords of the Admiralty. On the right and left of the royal stand were the stands of the House of Lords and House of Commons respectively, clad in green cloth. A little further down the sides of the slip, platforms were provided for naval and military officers, who were required to be in full-dress uniform, and for the town council and press. The roof of the slip and the columns supporting it were decorated with an accumulation of flags, as were the roads leading from the gate to the slip.

A little late, at 12.08 p.m., a royal salute of twenty-one guns from the *Duke of Wellington*, the *St Vincent* and the Garrison Battery indicated the approach of Princess Louise, whose train from London had been slightly delayed. After a religious service, the chief constructor, Mr W.B .Robinson, and the Mayor of Portsmouth were presented to the princess, and the mechanism for the launch was explained to her royal highness by Mr Robinson. This was followed by the christening and launching of the ship. A bottle of wine was dashed against the ram of the ship and the dogshores

The launch of the battleship *Inflexible* at Portsmouth on 27 April 1876 was a great state occasion, with royalty, the Lords of the Admiralty and representatives of both houses of Parliament present. Note the covered slip in which the hull had been built. The events of the day marked both the launching and the official opening of the Great Extension. (*Illustrated London News*, 6 May 1876)

of another twenty-one-gun salute. By permission of the Admiral Superintendent, the leading men engaged in the construction of the *Inflexible* dined together after the launch in their own hall, which had been 'tastefully decorated with flags and evergreens etc.'. A few speeches were made and the festivities were kept up during the afternoon. Meanwhile, the *Inflexible* was towed round to the new Tidal Basin, and would subsequently move to No. 13 Dock for fitting out.[71] Her name was appropriate: she was fitted with 24in side armour, the thickest ever mounted on a ship (and a thickness that was never exceeded subsequently). Her first captain, when she was commissioned in July 1881, was John 'Jackie' Fisher, who eventually became the First Sea Lord.[72]

Two additional, larger dry docks (Nos 14 and 15) were provided for in Clarke's original plans in case they were needed later, and the

holding the ship were knocked away. The ship did not move, so a dozen hands applied themselves to the hydraulic rams and at last she began to move, almost imperceptibly at first, gathering impetus and sliding evenly and gracefully down the slip, while the bands struck up *Hearts of Oak* and round after round of cheering was given. As the excitement subsided, the princess, accompanied by the Lords of the Admiralty, embarked in the royal yacht *Alberta*, which steamed through the Tidal Basin and then through the North Lock into the Repairing Basin, and disembarked at the sheers at the eastern end, where a temporary landing place had been constructed and draped with red cloth. The sheers were decorated with flags of all nations flying from the ordinary guys and from numbers of ropes reaching from their summit to the ground. A short train of five carriages took the princess and the accompanying party to the head of the new No. 13 Dock, where the princess declared the Great Extension open. A luncheon in Admiralty House followed, and afterwards the princess's departure was marked by the firing

The Dromedary-class dockyard paddle tugs were fine examples of late Victorian engineering. *Volcano* was launched by Barclay Curle, Glasgow, in 1899 and was based at Portsmouth. In 1918 she was renamed *Volatile*. 'A sight and sound to be remembered by those who knew (these tugs) was that of warming through the engines on a dark morning. The engine room was lit by strategically placed duck lamps with long flickering smoky flames, four huge cylinders gently rocking in the gloom, two steam reversing engines clattering around with water drains open – hissing steam.' (Hannan, *Fifty Years of Naval Tugs*, 1987, p.108) After many years' service, *Volatile* was scrapped in 1957 when replaced by the new diesel-electric paddle tug *Forceful*. (Portsmouth Royal Dockyard Historical Trust)

entrances to them were constructed when the Repairing Basin was excavated. In 1893, after a delay of twelve years, a start was made to complete them, and No. 14 was built with the original inverted arch entrance and width of 82ft. Both docks had a length of 563ft 6in, but the entrance to No. 15 was replaced with one of 94ft width and a squarer section, to accommodate the midship section of contemporary battleships, which had bilge keels to limit the rolling motion when at sea. A dam had to be built on the outside of the entrance to No. 15 Dock so that it could be dismantled and rebuilt to the new shape and size. Nearly 800 men were employed excavating the docks by both day and night, using Lucal lamps (which burned paraffin and petrol) for illumination at night. The bottom of the dock was made up of 3ft of concrete, 7ft of brickwork and 4ft 6in of granite blocks, giving a total thickness of 14ft 6in. The sides of the dock varied from 5–10ft thick. The docks were completed in 1896 and were big enough to take the late Victorian battleships: the Formidable and Duncan classes of 1898–1901 were 432ft in length, too long for the other Great Extension docks.[73]

Growth in Workforce Numbers

The new facilities would require big increases in workforce numbers. The table below shows workforce numbers in 1862–3, totalling 3,861 before work started on the Great Extension:

Numbers of Workers 1862–63

Officers

Superintendent	1
Master Attendant	1
Master Shipwright	1
Storekeeper	1
Accountant	1
Store Receiver	1
Timber Inspector	1
Assistant Master Attendant	1
Assistant Master Shipwright	2
Foremen of the Yard	7
Converters	2
Measurers	4
Inspector of Shipwrights	13
Boatswain	1
Master Smith	1
Master Rope Maker	1
Master Rigger	1
Master Sail Maker	1
Foreman of caulkers	1
Foreman of joiners	1
Foreman of smiths	2
Foreman of rope makers	1
Inspector of caulkers	2
Inspector of joiners	2
Engineer and machinist	1
Layers in the rope yards	2
Leading men of storehouses	4
Chaplain	1
Surgeon	1
Assistant Surgeon	1
Schoolmasters	2
Chief Engineer	1
Assistant to Chief Engineer	1
Asst Inspector of Steam Machinery	1
Foreman of Factory	1
Foreman of boiler makers	1
Timekeeper	1

Establishment of dockyard workmen

Shipwrights and apprentices	900
Caulkers and apprentices	86
Joiners and apprentices	148
Sawyers	92
Smiths and apprentices	199
Workmen and apprentices at millwright's shop	78
Workmen at block-, saw- and metal-mills	69
Riggers	260
Sail makers and apprentices	40
Spinners and house boys	120

Other trades and apprentices	125
Labourers	203
Total	**2,320**

Hired labourers, first-class	226

Establishment of Steam Factory Workmen:

Draughtsmen, writers, etc.	11
Asst timekeepers and collectors of work	3
Leading man of stores	1
Store porters, messenger boys, etc.	4
Engine keepers	10
Patternmakers including leading men, assistants and boys	30
Fitters and erectors, assistants and boys	170
Boiler makers, assistants and rivet boys	214
Engine smiths, hammer men, etc.	38
Founders	64
Coppersmiths	20
Millwrights	22
Painters	3
Labourers	160
Total	**750**

Hired artificers and labourers

Hired coppersmiths etc to make Grant's condensers	45
Hired spinners, labourers and house boys for getting up store of yarn, etc.	89
Colour women	7
Hired artificers and labourers with reference to the establishment of shipwrights	424
Total	**565**

Source: House of Commons Sessional Papers 1863, no. 191, in vol. xxxv; 1863, pp.1–3, pp.163–65 in vol.

In 1864–65, the Navy Estimates provided for 2,194 shipwrights, other artificers and labourers on the establishment and 226 first-class hired labourers. In addition, there were 2,012 hired shipwrights and other artificers and other hired labourers. The total workforce was thus 4,432. This included those employed in the Ropeyard, boys, and women employed in the Colour Loft. There was a reduction in hired employees over the next five years, to just 663 plus 200 first-class hired labourers. The total workforce had reduced to 3,815 by 1869–70.[74] By March 1876, the workforce had expanded again, to 4,910 (1,690 established and 3,220 hired).[75] Following the opening of the Great Extension, numbers rose to 7,727 by 1886,[76] though dropped to 7,024 by 1889 following a period of retrenchment which saw discharges of hired men. Such discharges were made towards the end of the financial year when the labour budget had been spent.[77]

The already tacit two-power standard was formalised by Lord George Hamilton, First Lord of the Admiralty, under the 1889 Naval Defence Act, which authorised increased spending to increase the size of the fleet as fears about the combined naval strength of France and Russia grew, and numbers of dockyard employees rose again. Under Earl Spencer's even more ambitious programme of 1895, the Navy's budget grew from an average of £5.02 million in 1889–94 to an average of £9.2 million in 1895–99, leading to a massive expansion of the dockyards' workforce.[78] By 1900, the workforce at Portsmouth had grown to a new high of 10,044.[79]

One imperative was to speed up the construction of new ships: the eight battleships that were laid down in the dockyards between 1880 and 1884 took, on average, five years and nine months to complete. After 1885, much larger numbers of men worked on each ship than previously, to maintain full and continuous production, and this – coupled with other measures such as improved procurement procedures – proved effective in reducing completion times without extra cost.[80] As the new arms race accelerated, Portsmouth became particularly busy with new construction in the1890s, including eight battleships and six cruisers.[81] The battleship *Royal Sovereign* was 'launched' on 26 February 1891 by Queen Victoria; the ceremony was actually associated with the floating out of the ship from No. 13 Dock, where she had been built, into the basin. She was completed in two years and eight months, four months ahead of schedule, spurred on by the then Admiral Superintendent, John 'Jackie' Fisher. It was the first time in over thirty years that Queen Victoria had attended the launch of a ship at Portsmouth, and it was a double ceremony because she also performed the naming of the cruiser *Royal Arthur* at her launching from the main slip on the

The battleship *Formidable* was built at Portsmouth and launched in 1898. She was sunk on New Year's Day 1915 by *U24* off Portland Bill.

same day. The queen was accompanied by her three surviving sons, and the German battleship *Oldenburg* was in attendance, with her admiral representing the German emperor.[82] In 1892, the battleship *Centurion* was launched, followed in 1895 by two battleships and one in each of the following four years.[83]

The increase in tonnage launched in each decade during the late Victorian and Edwardian eras is shown below:[84]

| 1871–80 | 55,720 | 1881–90 | 65,488 |
| 1891–1900 | 145,820 | 1901–10 | 152,950 |

The Diamond Jubilee Fleet Review at Spithead on 26 June 1897 saw what was then the most powerful force ever assembled by the Royal Navy: 165 flag-bedecked ships, including twenty-one battleships and forty-four cruisers, all drawn from home waters, with none withdrawn from foreign and colonial stations. The ageing monarch was too old and frail to be present, and instead it was the Prince of Wales who reviewed the fleet from the royal yacht *Victoria and Albert* whilst the queen watched from the grounds of Osborne House, her favourite retreat, on the Isle of Wight. *The Times* recorded the scene:

Judged as a pageant nothing could be more impressive than the long lines of ships anchored in perfect order, spreading over miles of water in apparently endless array. Each vessel, in its bright setting of sea and sky, forms a picture in which there are no harsh tones. The pale yellow funnels, white upper works, and slender masts relieve the somber low-lying hulls. Human interest is abundantly provided by the fringes of bluejackets cheering heartily as the royal yacht passes slowly down the lines. The splendid panorama is all in perfect harmony, and one almost forgets the vast forces which are latent in this peaceful assemblage of warships.[85]

The old queen died at Osborne House on 22 January 1901, and on 1 February her coffin was taken by the royal yacht *Alberta* to Portsmouth harbour to be conveyed by train from Clarence Yard to Windsor.[86]

From Apprentice to Chief Constructor and Wooden Walls to Dreadnoughts

James A. Yates was born in 1852, and from 1863–66 attended the Royal Hospital School at Greenwich, where there was a fully rigged ship in the centre of the grounds, on which the older boys were trained and exercised. The school building is now the National Maritime Museum. In 1866, Yates passed out as head of the school and entered Portsmouth Dockyard as an apprentice shipwright at the age of 14. In the early period of his apprenticeship wooden shipbuilding was passing away, but two wooden sloops, *Sirius* and *Dido*, and a number of wooden gunboats were being constructed at Portsmouth. He recalled the work on the curved frame timbers:

A number of clean and intelligent shipwrights, each with a rule, a bevil, calipers and compasses on his person, the ground covered with fragrant chips, the cheerful sound of the dubbing adzes, the wholesome scent of the Stockholm tar with which all fitting surfaces were coated, and the sight of the timbers of the growing ship, formed altogether a very pleasant experience.

He described the planks being fitted:

After trimming, the plank was put in a steam kiln for several hours, and it was a peculiar sight to see the steaming plank being carried to the ship on the shoulders of the men. Another inspiring experience was the sight and sound of hundreds of caulkers caulking the sides of a great ship with their shrill musical mallets.

However, the wood was rapidly being replaced by iron, and 'shipwrights became drillers, platers, riveters and iron caulkers'.

In 1871, Yates was selected for training as a naval architect in rooms attached to the South Kensington Museum, a school which moved to the Royal Naval College at Greenwich in 1873. In 1875, he became a draughtsman at the Admiralty, the start of a career which led to him being appointed, by Sir John Fisher, as Chief Constructor at Chatham in March 1893, and subsequently, in early 1895, Fisher arranged his transfer to Portsmouth as Chief Constructor. He held this post for seven years, during which he launched or floated out eleven new ships, including six battleships. These were strenuous years. Other senior Admiralty roles followed until his retirement in 1910, but he was recalled for further service in connection with the war, between 1916 and 1918. He died in 1941 at the age of 89.[87]

10

THE DOCKYARD APPRENTICE

'among the problems ... engaging the attention of employers, education authorities, and teachers, few can compare in importance and extent with those relating to the proper training of industrial workers and to the provision of suitable means for the development and advancement of workmen of special ability. These objects have been attained with conspicuous success by the Admiralty Scheme.'

Board of Education, The Admiralty Method of *Training Dockyard Apprentices*, August 1916

An institution at the heart of the dockyard was the apprenticeship. Competition for entry to the apprenticeship scheme was intense, this in the time before many boys progressed to A-level or higher education, and some local private schools and evening classes provided special preparation for the entrance examinations, as did the municipal schools in the dockyard area. As well as providing qualified craftsmen for the dockyards, the scheme allowed the brightest apprentices to embark on a professional career in which they could reach the highest levels of the engineering profession.

The principal trade in dockyards during the sailing Navy era was the shipwright, and in Henry VIII's period, when there was a need to build a ship or fit out a fleet, shipwrights were often impressed from the multitude of small private shipyards around the coast, to augment the nucleus of staff who had more permanent employment in the dockyards, and this practice appears to have continued at least until 1708. The shipwrights brought with them apprentices, who were known as their servants, and in these periods of emergency the authorities were quite prepared to pay for them. In the dockyards in 1548, the principal shipwrights were expected 'to instruct others in their feats', but their 'servants' were the future dockyard officers, who were to gain a knowledge of ship design. The training of the ordinary apprentice was left to the craftsman to whom he happened to be indentured. In the time of the Commonwealth, the manning of the dockyards was placed on a more permanent footing, and in 1664 the Navy Board issued an order that every apprentice was to be aged 16 at the time of entry and was to serve seven years. Permission was given to enter a servant 'unto any person we find able and diligent', and the system of patronage meant that the apprentice had to give the whole of his wages to the officer or man to whom he was bound. The time and expense incurred meant that children of the poorer classes were effectively excluded, often being expected to do something for themselves from the age of 5. A boy whose father was already in the trade was in a much more advantageous position, for being apprenticed to his father would make a real addition to the family income. No test was imposed on entrants, except for age and stature, 4ft 10in being the minimum standard.[1]

This continued in the eighteenth century, with apprentices recruited from the well-educated boys of good families and allocated to a dockyard officer or, less frequently, a shipwright, but the instruction they received was often indifferent. Most of the apprentice's wages still went to the dockyard officer, the system still being based on patronage.[2] A dual system of superior and inferior apprentices was approved in 1800, but the proposed school for the education of the superior shipwrights was not established. The source of the superior apprentices dried up, and even many working-class parents no longer apprenticed their sons. From 1801, only two-thirds of the apprentice's income was given to the instructor and one-third went to the apprentice's parents or guardians. Thereafter, apprentices were mostly recruited from 'the lowest class of people, and those least likely to have any education'. In 1805, the Admiralty was persuaded by Samuel Bentham that deserving shipwright apprentices should be articled to the master shipwright, and assigned by him to a qualified shipwright instructor, alongside whom the apprentice would work. But the instructors were poorly supervised and the system did not work very well.[3]

Before 1843, when the dockyard schools were established by the Admiralty, little or no technical education had been provided in the dockyard towns. An exception was the School of Naval Architecture

Men leaving Unicorn Gate, c.1910.

at Portsmouth, which was open from 1811–34 and took a maximum number of twenty-four students on a seven-year apprenticeship. The entrance exam consisted of Euclid, arithmetic (and 'algebra to quadratic equations'), English grammar and dictation, and reading and translating French, and was for boys of 'gentlemanly stock'. Apprentices were lodged, boarded and educated free, and were paid salaries of from £25 to £60 in lieu of clothing etc. On completion of the apprenticeship, they were 'eligible to all situations in the Shipbuilding Department of His Majesty's Service', provided they had 'completed the plan of education and been certified by the Professor to be properly qualified'.[4]

This school closed following a change of government, but the pressing need for apprentice education soon led to a new initiative. A dockyard school for apprentices was opened in each of the home dockyards in 1843–44. At first the apprentices received a 'most elementary' form of education, covering basic education rather than technical subjects, each afternoon for three hours, and the apprentice-ship was for seven years.[5] The curriculum included numeracy, literacy and religious education: part of the Admiralty's motivation in opening the schools was to instil a sense of discipline and moral rectitude amongst that part of their workforce that was drawn from the lower working classes, as well as improving skill levels.

The combination of off-job education and training with on-job training was innovative and acted as a precedent for many later sandwich education schemes.[6] Initially, only one teacher per school was employed (being appointed from the dockyard officers) and the dockyard schools were unpopular not only with apprentices but also with their teachers, 'who were often reluctant volunteers prone to becoming more reluctant once they had experienced the anarchy inherent in supervising over two hundred caged youths'. A series of reforms followed: more teachers were appointed and the status of the schools was improved.[7]

In 1846, the Admiralty appointed a senior inspector for education, and this led to the splitting of the intake by merit into Upper and Lower Schools, the latter being for the minor trades. After the first three years of the apprenticeship, the best apprentices were 'creamed off' for a technically more complicated curriculum designed to breed individuals for the 'inferior' officer positions and, eventually for a further minority, the 'superior' management posts. From 1848–53, this group was trained at the Central Mathematical

School at Portsmouth. In 1864, the Royal School of Architecture was opened at Kensington, and soon moved to Greenwich, taking the upper tier of apprentices after their third year in the dockyard school.[8]

The curriculum was gradually extended by the inclusion of subjects such as mathematics, the elements of natural philosophy, applied maths, mechanical drawing, physics, applied mechanics, heat engines, metallurgy, naval architecture and electrical engineering.[9] These subjects were increasingly needed as the Navy moved from wood and sail to iron, steel, steam and electricity. In 1873, new regulations for the entry of apprentices were issued by the Admiralty. They specified that boys sitting the entrance exam must be aged 14 or 15 and would normally have to meet the minimum requirements for height (4ft 8in), weight (90lb) and girth of chest (26in), though exceptions might be made if the dockyard Medical Officer thought the candidate would make an efficient workman. They would sit a preliminary exam in arithmetic, orthography, handwriting and grammar, and if they passed that would sit a further exam in English composition, geography, Euclid (first three books) and algebra, including quadratic equations.

It was a seven-year apprenticeship and the pay in the first year was 3s per week, rising by 2s each year so that in the seventh year it was 15s per week. A relative or friend had to be 'able and willing to undertake the duty of the second party to the indenture as to the support etc of the Apprentice during his apprenticeship'. Board wages would only be paid for 'boys whose fathers have died or been killed in the service, and when it can be proved the family are in distressed circumstances'. Three boys who passed five years would be selected annually by competitive examination for study at the Royal Naval College at Greenwich for one year, followed by a year at sea in a Royal Navy ship or in a situation at a dockyard. They would then be appointed Assistants to the Foremen of the Dockyards or any other post for which they might be considered fit.[10] Until 1878, the full benefits of the school were confined to shipwright apprentices, but thereafter the minor trade apprentices could enter the Lower School, and from 1893 all apprentices could enter the Upper School if their examination results warranted it.[11] The length of the apprenticeship was reduced to six years in 1890, and to five years in 1918. The age on entry was lowered to 15 in 1765 and to 14 in 1769, but was raised to 15 again in the First World War.[12]

Sir William White, Director of Naval Construction from 1885–1901, himself a former Devonport Dockyard apprentice and always closely involved with dockyard education, spoke of the wider benefits of the apprenticeship: 'There have been hundreds of young men … who have been helped to fight evil habits, whose minds and bodies have become wholesomely occupied, who have been saved from becoming hooligans, and made better men.'[13]

Dockyard Schools in the Twentieth Century

In 1905, laboratories for practical science work were introduced and the whole scheme was redrafted to make it more rigorous and practical in nature. At this time, the apprentices attended for twelve hours each week over what was by then the six-year period of their apprenticeship. There were six dockyard schools in 1916 – at Portsmouth, Chatham, Devonport, Sheerness, Pembroke and Haulbowline – and another was about to be opened in the new dockyard at Rosyth. Entry was by a competitive examination held annually in London, Edinburgh and each dockyard town. It was open to all boys of British nationality aged between 15 and 16. In practice, however, it was confined to boys whose parents or guardians lived within reach of the dockyard, as the boys, if successful, had to live near their work. The exam was conducted by the Civil Service Commissioners and covered the subjects of English (composition, literature and handwriting), history and geography, arithmetic, algebra and geometry, drawing and elementary science. The different trades covered by the apprenticeships in 1916 were shipwrights, engine fitters, electrical fitters, ship fitters, patternmakers, boilermakers, caulkers, coppersmiths, joiners, founders, plumbers, painters, ropemakers, sailmakers and smiths. Each successful candidate was able to choose from the trades in which there remained vacancies after each boy above him (in the results table) had made his choice. The examination also served for entry of boy artificers and boy shipwrights into the Royal Navy.

The best-performing candidates in the examination were allocated to the Upper School and attended on two afternoons and three evenings each week for about forty weeks each year, with twelve hours per week, over four years. The others attended

Men leaving the battleship *Bellerophon*, which was fitting out from 1907–09. (Portsmouth Royal Dockyard Historical Trust)

the Lower School on one afternoon and two evenings each week, with seven hours per week, over three years. The curriculum for the Upper School in 1916 was as follows:

Year 1: English, Practical Mathematics, Elementary Science, Mechanics.

Year 2: English, Practical Mathematics, Mechanics, Heat and Metallurgy, Mechanical Drawing.

Year 3: Practical Mathematics, Applied Mechanics, Electricity, Mechanical Drawing; plus Steam and Heat Engines for engine fitters and kindred trades and electrical fitters; General Engineering for engine fitters and kindred trades; Naval Architecture for shipwrights and kindred trades.

Year 4: Practical Mathematics, Applied Mechanics, Electricity, Mechanical Drawing, Heat and Metallurgy; plus Steam and Heat Engines and General Engineering for engine fitters and kindred trades; Electrical Engineering for electrical fitters; Naval Architecture for shipwrights and kindred trades.

When not at school, the apprentices were placed with a workman of good character to learn the practical side of their trade, and also spent a period in the drawing office. The final two years of the apprenticeship were also spent in this way.

'Idleness or indifference' brought summary dismissal from the school. At the end of each year, those failing examinations were dismissed from the school (or might be demoted from the Upper to the Lower School), whilst the best students in the Lower School might be promoted to the Upper School. The number rejected at the end of each year was about half of the total. Those rejected from the school had to rely on the local technical school for instruction.

Shipwright apprentices who completed four years in the Upper School with especially good results in the final examination were recommended for Cadetships in Naval Construction, and, if appointed, were given a year's course of combined theoretical and practical instruction in the dockyard, or later at the Royal Naval Engineering College, Keyham (subsequently at Manadon, Plymouth), before proceeding to the Royal Naval College at Greenwich for a further course of three years. On successfully achieving a first- or second-class certificate in their final examination, they were confirmed in the Corps of Naval Constructors. In a similar way, electrical fitters could qualify to become Assistant Electrical

Engineers. Those performing less well in the final examinations were eligible to become draughtsmen or inspectors.

The ability to progress in this way to professional posts was a valued part of the dockyard apprenticeship scheme, and some former apprentices reached the highest level of their profession. Sir Philip Watts began his Portsmouth apprenticeship in 1860 and qualified as a naval constructor in 1870. In 1885, he left the Admiralty service to become Naval Architect and General Manager of Sir W.G. Armstrong & Co., where he designed and built warships for the Royal Navy and foreign navies. In 1902, he returned to the Admiralty to become Director of Naval Construction and headed the team which designed HMS *Dreadnought* and the successive classes of Dreadnoughts until his retirement in 1912.[14]

The Admiralty did not always find it easy to recruit enough boys of the right quality, a problem that was exacerbated in wartime when boys might prefer to join the armed forces. In 1941, there were 354 apprentice positions available at Portsmouth, plus eighteen naval shipwrights, and candidates had to be aged 15 or 16 when they sat the examination.[15] Only 217 of the posts were filled, leaving a deficit of 137. Some of this deficit was filled by yard boys, who left school at age 14 to take unskilled dockyard jobs. A particular problem was that the job of shipwright had become very unpopular with boys and their parents, and fifty-five of the 100 shipwright apprentice vacancies had to be filled by yard boys. This contrasted with 1940, when only three of ninety-eight shipwright apprenticeship vacancies needed to be filled by yard boys. An interesting fact was that 63 per cent of the boys in the top twenty examination results positions were at Portsmouth, with the other three dockyards (Devonport, Chatham and Sheerness) contributing the remaining 37 per cent. An explanation for this may have been the 'crammer' schools in Portsmouth, such as the Mile End School (known as Oliver's) which specialised in the apprentice exam. One boy who took the exam in 1943 described that school as 'brutal, teachers used canes of various sizes and any mistake or fault was punished on the spot with whacks across the hands or one's back made with some ferocity … Oliver's was a school where "success" and "thrash" were synonymous.' Of the top twenty entrants at Portsmouth in 1941, only two chose the shipwright apprenticeship, and this pattern was replicated at the other dockyards. Many reasons for the unpopularity of the shipwright position were

advanced, including the very extensive knowledge which had to be learned, the fact that the skills were less transferable outside the dockyard when compared with many other trades and the history of the wholesale discharge of shipwrights after the previous war. The electrical trades had become the most popular apprenticeships.[16] Another problem was that retention rates were alarmingly low: of the 323 apprentices recruited at Portsmouth in 1940, only 121 (37.5 per cent) moved into the second year and the remaining 202 left. This problem was not confined to Portsmouth, since during the war years approximately 900 apprentices were recruited each year to the five dockyard schools (Rosyth had reopened), but less than 100 remained by the fourth year.[17]

In 1941, the school buildings in Portsmouth Dockyard were destroyed in an air raid, and temporary accommodation was found at the Teacher Training College in Milton and in other buildings nearby. In 1945, the school was housed in Nissen huts outside the Unicorn Gate, and from 1950 in a building at Flathouse vacated by the Mechanical Repair Establishment, which moved to HMS *Sultan* in Gosport. In 1952, the title of the Dockyard School was changed to the Dockyard Technical College. In the 1930s, the length of the apprenticeship was reduced to five years, and to four years in 1967. From 1956, the National Dockyard Entry Examination was dropped and a dual-entry scheme was introduced – comprising Student Apprentices who were trained to become inspectors or draughtsmen, and Craft Apprentices who were trained as craftsmen. In the 1960s, technician apprentices were introduced, replacing student apprentices, and their five-year scheme allowed them to take up posts as draughtsmen or technical grade officers. The Dockyard Technical College now awarded Ordinary and Higher National Certificates (ONC and HNC) in Naval Architecture, and Mechanical and Electrical Engineering. The number of trades reduced sharply in the second half of the twentieth century through amalgamation and increased flexibility, and female apprentices were recruited into some of the craft trades from 1969.[18] In 1970, the Dockyard Technical College closed and the education activities were taken over by Portsmouth Education Authority, still using the Flathouse building.[19]

The intake of apprentices into Royal Dockyards ceased in 1981, but in 1984 recommenced on a limited basis in a single-stream entry under the private contractor responsible for the operation of each dockyard or naval base, which at Portsmouth was the Fleet Repair and Maintenance Organisation. Initial 'off-job training' and academic studies was undertaken at Highbury College until 1998.[20]

An Apprentice in the Great War[21]

Edward Lane was born in Margate Road, in the Somers Town district of Portsmouth, on 2 January 1902. In 1913, he passed an examination to attend the new junior technical school for boys at a cost to his parents of £3 7s 6d per year. In 1915, he passed the dockyard apprentice examination at his first attempt and was persuaded by his parents to leave the school, entering employment in the yard on 12 July 1915, aged only 13. He wept when he left the school, which he loved. For working-class people in Portsmouth, though, the great achievement for a boy was passing the dockyard examination, and it was the parents who decided on their children's careers. But Lane later thought that he would have benefited from a further year at school before entering the dockyard, as it would have placed him on a better footing with his peers and his performance would have been better.

It was wartime and the dockyard was a hive of activity, with the battleship *Royal Sovereign* just launched and fitting out. In No. 15 Dock, three K-class submarines were being built, whilst two J-class submarines were nearing completion in No. 13 Dock. He was taken by his chargeman and handed over to his instructor, a craftsman who received an extra two shillings a week for the honour. Apprentices were known to their instructors as their 'Two Bob', and one instructor would say to another, 'I wonder what my Two Bob is up to now?' At first Lane was completely overwhelmed by his surroundings and the company he was in:

> The language and the topics of conversation were in the main disgusting and my instructor had nothing to do with this. He was a loyal member of the union and told me not to listen to the ribaldry and to keep well away from that sort of company.

As he approached the Unicorn Gate in the morning, the bell would be tolled by one of the policemen, starting at 6.45 a.m. At 6.55 a.m. it would start tolling twice as fast, and often Lane only

This 31ft steam cutter, built in 1896 by J. Samuel White, Cowes, was carried aboard successive royal yachts until 1936, and is shown on display in No. 4 Boathouse, where she is under the care of Portsmouth Naval Base Property Trust.

got in by putting his foot in the door of the muster station just as the timekeepers were closing it and giving it a mighty shove with his shoulder, nearly pushing the timekeeper over. If he did not reach his work place by 7 a.m., he would lose one-sixth of a day's pay (unless he was working with a 'good sort' of chargeman), and unpunctuality could lead to a day's suspension. The day's work was from 7 a.m. to midday, without a break, then 1.30–5 p.m.. On Saturdays it was 7 a.m. to midday. At 4.15 p.m. on Fridays, they queued at the pay office for their pay envelopes, and the many pubs just outside the dockyard wall did a brisk business as the men left the yard. During the years of the Great War, overtime was continuously worked, including Sundays and the few public holidays.

At the top of the dockyard hierarchy was the Admiral Superintendent, a job that Lane said was a sinecure. Apprentices had a lot of homework, and if not needed by their instructors they would hide away behind the tool boxes in a tool shed to do this work. On one occasion, Admiral Weymouth, the Admiral Superintendent of the yard, was walking by and saw a boy bob up. Lane recalled:

Now Admiral Weymouth was a nasty old man and always ferreting about for anyone he thought was idling. He found a chargeman and sent him over to get our names and numbers and we were hauled up by the Manager Construction Department and given a dressing down.

The young Edward Lane had been good at woodwork at school and did well enough in the dockyard entry examination to be able to choose to become a shipwright apprentice. Shipwrights were at that time the elite of the dockyard craftsmen, and had historically been the dominant trade; in the days of wooden ships, they had constructed them. When iron replaced wood, the boilermakers took over construction in the private shipyards, but in the royal dockyards the shipwrights maintained their hold on construction. During 1917 and 1918, Lane worked in No. 1 Shipbuilding Shop, near the Marlborough Gate, where light plate work was done:

The noise from the place was deafening, huge punching and shearing machines and hydraulic presses operated by skilled labourers. The shipwrights set out and planned the work for the machinists and then assembled the work ready for fitting on the ship … The work was mainly done on piecework and there were many arguments over the price to be paid.

Lane took a dislike to his instructor there. Whilst conceding that he was a very ingenious and capable craftsman, he thought him to be crafty and greedy: 'He was an established man, and a Tory of course, and lost no opportunity to condemn anything done by the union.'

All shipwrights and apprentices entered the yard as hired men with no right to permanent work. Vacancies amongst the established men, who had job security and a small pension, arose by death or retirement, which was at the age of 60. The established man was 'alright for life provided he did nothing silly'. On becoming established, 2s per week was deducted from pay to go towards the pension. Many men dropped out of the union to make up for the loss, what Lane called a case of 'I'm alright Jack'. The Workers' Hospital Maintenance Fund was inaugurated during the 1920s. It was voluntary, but most workers joined and for a small weekly 'sub' they were entitled to free hospital treatment.

Lane was glad when his instructor arranged his removal from No. 1 Shop, and Lane got away from the 'money grabbing of piece work'.

Still an apprentice, he worked on the preparation of the battlecruiser *Renown* for the Prince of Wales' overseas tour. Built during the war, the ship had no luxuries and was to have a new teak deck. It had to be done quickly, so the shipwrights worked day and night shifts, laying hundreds of tons of teak in 9in x 3in planks. The resulting rough surface was then trimmed with an adze to a fair surface, using the adze across the grain, before being machine-planed, though in the fiddly places it had to be hand-planed. When it was finished, Lane thought it looked a beautiful piece of work. On qualifying as a shipwright, he elected to work in one of the boathouses where ships' boats were repaired. One of his jobs was to fix the coat of arms ('a beautiful specimen of the signwriter's art') on the port and starboard sides near the bow on the royal barge, which was kept in the boathouse when the royal yacht *Victoria and Albert* was laid up off Whale Island. The barge was the pride of the boathouse. '[T]he best painters would prepare the hull with hard stopping and about six coats of paint rubbed down after each coat to obtain a mirror like finish,'' and with 'gold leaf in small sheets they applied the gold ribbon around the topsides. All the fittings – funnel, fairleads, bollards – were silver plated.'

Building the big battleships was enormously labour intensive, with everyone getting in everyone else's way. 'The noise was deafening and accidents frequent. The dockyard dealt with minor accidents, serious ones were taken by boat across the harbour to the naval hospital at Haslar.' One of his colleagues was killed when, as he was standing on staging in a dry dock, a steel plate fell on him as it was being lowered by a crane towards a battleship which was having bulges fitted to its hull. On one occasion, Lane fell head-first into the boat pond whilst manoeuvring a steam pinnace into position for slipping. 'I had to swim for it. Fully clothed, overall and all. My long boathook had slipped and in I went. I was taken straight to the surgery and a small van took me home.'

In 1928, Edward Lane passed an examination to become an inspector, but at that time opportunities for promotion into an inspector's post were few so he decided to look around for other opportunities. In 1930, Lane left the dockyard to become a school teacher at Clarence Square Senior School in Gosport, having taken evening classes to obtain a City and Guilds qualification for teachers of handicraft. He was recalled into the dockyard in 1941 and worked in the drawing office, before becoming an inspector of shipwrights. In 1945, with the war over, he left again and became the woodwork teacher at Gosport County Grammar School, where the writer was one of his pupils.

Life as an Apprentice in the Twentieth Century

The first hurdle for the aspiring apprentice was the entrance exam, and to prepare themselves for it boys often attended a cramming school such as Chivers or Olivers, or a school such as St Luke's Church of England School, which had a high reputation for getting boys through the exam. A retired shipwright who started his apprenticeship in 1937 recalled:

I went to St Luke's School and then entered the Dockyard School. St Luke's had a class where we were schooled extensively for eighteen months to take the dockyard examination. We had about ten papers … it was a three-day examination … We had to read a book and had

Dockyardmen attend election hustings in December 1910. At this general election, the Liberal Party gained power and in 1911 introduced a national insurance scheme which gave workers sickness and unemployment benefits in return for contributions from both employees and employers.

a paper on that, then there was maths, trigonometry, algebra, then science, geography, history and a handwriting paper.[22]

A former shipwright apprentice who started in 1956 remembered:

There were two forms of entry, into the … dockyard apprentice scheme, one was by an aptitude test and an interview which took you into the Lower School, the other was by open examination which was … nationwide, for all the dockyards, and on passing that, you went into the Upper School of the dockyard where you were put on an Ordinary National Certificate course, then if you were lucky enough to complete that, a Higher National and so on, and I think eventually, you got accredited with a BSc.[23]

Depending on their ranking in the pass list, successful candidates were either given a choice of trade (if they were high up on the list) or offered particular trades. One former joiner apprentice who started in 1959 said, 'I came 112th. There was a cut off line … sort of every 20 … because I came between 100 and 120 there was a list of jobs allocated for that twenty … I went in as a Bench Joiner and Cabinet Maker.'[24] A retired locksmith recalled:

I took a two-day exam to get into an apprenticeship in the dockyard in 1904. The upper-crust ones, they took the better jobs and we had the rest! I went into the Smiths' Shop [No. 1 Smithery] … The shop itself was as big as two churches put together and as tall … it was part of the old Crystal Palace brought into the dockyard after the 1851 exhibition … it was a cast-iron structure.[25]

For a young boy who had passed the entrance exam, the first days of his apprenticeship were often fraught as he tried to adapt to the new environment. An engine fitter who retired after forty-five years in the yard recalled:

I left school in August 1940. I left on the Friday and by the Monday morning I was in the dockyard. A very abrupt transition from being a schoolboy to being something else … and that something else was quite a shock. My life changed absolutely completely … I found myself in an environment that is beyond belief, unless you know what the conditions are like you wouldn't appreciate it. Because it was wartime, my first month was spent afloat on the battleship HMS *Queen Elizabeth* … It was a dirty, noisy, extremely dark place and you were surrounded by all sorts of alien things and rough workmen. As a schoolboy it's almost like leaving heaven and entering hell.

Another engine fitter apprentice recollected:

I think I was nearly in tears at the end of the day … it was very foreign to me, I was within these walls … everything was strange and different, and I just wanted to get home and be home again … by the second day they started giving us tasks to do, so you were sort of occupied then.[26]

The new apprentice was assigned to an instructor. 'With all the instructors, if you were really interested in learning, they would go to no end of trouble to teach you, nothing was too much trouble … if you weren't interested, they didn't want to know you either,' said a former engine fitter apprentice. A retired shipwright who had served forty-eight years in the yard recollected:[27]

If you wanted to hide away to study or read books, they [the instructors] would find somewhere for us to go and then make sure if the foreman came round he didn't see us … But it was all a game, because the foreman knew as well as anybody else … they'd probably done the same when they were young!

D.K. Muir, a former electrical fitter apprentice who started his apprenticeship at Portsmouth in 1937, remembers that the working week was forty-seven-and-a-half hours, excluding the lunch breaks, a five-and-a-half-day week, plus three nights at 'Tech' finishing at 9.30 p.m. – no day release; he had only one week holiday per year, this being in August during Navy Week. He cycled 12 miles each way to work and back in summer, to save money:

On passing my dockyard exam it was almost symbolic that my parents bought me a new bicycle. Virtually no tradesman owned a car, with few even affording public transport to work. Neither could they afford the luxury of canteen lunches, the main meal of the day, and to allow them to travel home for this purpose the lunch break was an exceptional 1½ hours. As a result, the sight of thousands of cyclists streaming to and from the dockyard four times a day was a scene never to be forgotten, and even there a class system held sway. Clerical workers were allowed to cycle through the dockyard gates to their place of employment where parking sheds were provided, whereas the lowly tradesman was forced to leave his machine outside the gate. This created a thriving business of cycle minding in small back yards with the worker then having to walk up to a mile to his place of employment before clocking in.

He remembers his instructors with some fondness:

Without exception our Instructors were intelligent, articulate and gentle men, who treated their charges with fatherly care, provided us with morning tea at their own cost, and above all, passed on their knowledge and skills as one would pass on secret religious rites.[28]

Every apprentice had to attend the Dockyard School. A retired dockyard school master who taught there from 1937–74 recalled:

It was the only way of getting to university in Portsmouth unless your parents could help and could afford it. The standard of education was incredible for a school in Portsmouth … It was degree standard. But it was really hard work and many boys couldn't put up with working all day and attending school up to eight in the evening after a day's work … All the staff were of degree standard … [they] had to teach other subjects than their own special subject and the classes combined for subjects such as history, so there would be about 60–70 in the class. The main part was engineering subjects … mathematics, mechanics, thermodynamics, engineering drawing … which we all had to teach.[29]

A former shipwright apprenticed in the 1950s remembered:

We didn't get paid for night school, so there was always the occasion to duck out of it …. They would come round and inform the inspectors … that you didn't attend night school. The inspectors would then inform your instructor and the instructor would invariably give you a rollicking and a clip round the ear and say 'I'm not going to lose my instruction money because you're not doing your job, so you get to night school.'[30]

The newly built battleship *Queen Elizabeth* runs sea trials in 1915.

In later years, all of the first year was spent at the training centre. One engine fitter/turner apprentice in the 1990s recalled, 'The training stuff we were given … it was like text book stuff, and sometimes, as you know the fitters outside, they would have learnt a different way, an easier way, and they would pass that on to you.'[31]
D.K. Muir recalled:

On joining the dockyard every apprentice was given a large wooden chest about 3ft long and some 18in square with rope lifting handles at each end. These were essential when working 'afloat' since the vessel could be an empty shell devoid of any amenities and the tool chest as well as being a safe repository for all your tools contained all your worldly goods, not to mention acting as a seat.[32]

Practical experience on the job depended on the trade, and the boys were often impressed by the workshops in which they found themselves in their first year. Joiner apprentices worked in the Joiners' Shop, which a joiner apprentice whose apprenticeship was during the Second World War remembered was:

probably the biggest joiners' shop and machine floor in the south of England … when I went in there was something like 150 joiners, French polishers, upholsters, wheelwrights. My instructor was Charlie Sneed … he was the best craftsman that I ever met in my life, he could make tools talk. He taught me a pride that I never knew existed … He said, 'Never forget you're a craftsman, you're not a fitter or an electrician, you can work directly from nature and that makes you a craftsman.'

An engine fitter who was also apprenticed during the Second World War recalled:

The first year was spent in the Factory doing bench work. The Factory was the biggest manufacturing building on the whole of the south coast. You made jigs, tools and gauges and it was precision work. At the end of the first year you did a trade test and you were also assessed on your merit … your character … and all that went into an annual report to the Chargeman of Apprentices and it went into your records.

A painter manikin in the Dockyard Apprentice exhibition in No. 7 Boathouse.

He progressed through more complicated bench work in his second and third years. It wasn't until his fourth and fifth years that he was sent out into the 'real world':

> We went afloat, to where ships were being refitted. In the same way you went from section to section, so you went from big to little ships – battleships, aircraft carriers, destroyers and frigates, minesweepers, submarines, fast patrol boats, gunnery sections, landing craft, tugs … it was engine room work and gunnery, gun mountings. In the fifth year I spent three months in the Drawing Office, finding about how drawings were prepared and about how trials were carried out on ships … You mostly worked on your own in the fifth year, or with another fitter.[33]

Apprentices were given practical trade tests. A former joiner apprentice who started in 1959 said:

> [E]very six months you would probably have … a job brought in and that would be your trade test … it was all assessed by people who came from other parts of the dockyard and they had a good look at it and then you were assessed on what they thought you were worth

as a … tradesman … [if you failed the tests] you did it again, 'til you got it right … you didn't fail unless you made a complete and utter bodge of it.[34]

He reflected on the whole experience:

> It was brilliant … the best apprenticeship any nipper could have had … it was hard … you were hammered, everything was hammered into you … you weren't allowed to get away with anything, but at the end of five or six years there was [sic] some absolutely brilliant tradesmen come out of Portsmouth Dockyard.[35]

But the transition from apprentice to a journeyman did not mark an end but only a stage in gaining experience. One former shipwright apprentice who passed out in 1959 recalled being amongst a group congratulated by the Senior Personnel Manager as he gave them their indentures:

> [Then] we walked along the road and bumped into one of our instructors from the training establishment … We proudly showed him our Indentures … he said, 'You might be shipwrights but you're not craftsmen … you're journeymen … you've got to journey round to get experience, and in seven years you might be of use to some poor bugger!'[36]

In 1949, the Admiralty decided to close the dockyard in Bermuda and the apprentices there had to come to England in 1950 to complete their training. One of the Bermudan apprentices remembers that:

> [M]ost of the apprentices were excited about coming to England … it was the first time that those boys had left Bermuda … you heard so much about England at school, but it was disappointing when you got there! We got off the boat at Liverpool … when we got to Portsmouth it was pouring with rain, and if I could have gone home that night, I would have gone.

A retired shipwright recalled:

> [A]bout thirty coloured boys arrived here … they were all trades and we had about a dozen shipwrights. Of course they were black and

The former School of Naval Architecture, completed in 1817, faces The Green on the opposite side to Admiralty House. It trained a superior class of apprentices capable of rapid promotion to dockyard officers.

they were a novelty. There was one bloke who used to ride home for lunch, Wesley Deal, and the kids used to follow him on his bike because they'd never seen a black man before! But there wasn't any racial prejudice then and we accepted them just as being apprentices. They had all finished their apprenticeships by 1955 ... some went back home, some married girls here and some stayed when they came out of their time, but they gradually all went back.

One of the shipwright apprentices who spent five years at Portsmouth later recalled, 'When I first went back to Bermuda I felt like a foreigner ... because I really grew up here, and I felt strange for a while. I was sorry to leave Portsmouth.'[37]

Following the Sex Discrimination Act, in 1969 the Admiralty agreed to the admission of female apprentices, though the numbers turned out to be small: between 1970 and 1980, twenty-nine were taken on. One preferred job for them was electrical fitter, which was believed to be a suitable trade for women, and twenty of the twenty-nine were apprenticed in this trade, and two as sailmakers. Heavier jobs were sometimes problematic. A retired shipwright said:

> We had one girl who worked in the Foundry [as an apprentice] ... but the complaints were that she couldn't lift the heavy ladles [for pouring molten metal into moulds]. You find that most girls don't want to come into the dockyard because of the environment.

The young women often found it difficult to integrate with the 'lads' at first. An engine fitter/turner apprentice who started in 1995 recalled:

> Well there was only me and one other female, and there was loads and loads of young lads ... You weren't too sure what to expect, how they were going to accept you. You were worried whether they would all start hating you and sort of bullying you in certain ways. It was very very scary. I was very nervous, you sat by yourself most of the time ... because if you sat down none of the lads would actually sit next to you in case the other lads would actually start teasing them about sitting next to one of the girls.

But, 'by the end of the four months everyone was relaxed and then the dirty jokes started coming out', and she would reciprocate by telling sexy jokes as well. 'They used to laugh 'cos when I used to say them, I used to go red and blush!'[38]

We want eight and we won't wait!

Tory MPs fuelling the public clamour for more
Dreadnoughts to be built each year, 1909

In 1900, Britain's naval supremacy was undisputed – the fleet was at least as big as the next two largest fleets put together (in accord with policy of the two-power standard). The potential enemies had been France, Russia and Japan, but new treaties reduced these threats, only to be replaced by an emergent one from Germany. In 1904, warning against complacency, the Second Sea Lord, Admiral Sir John 'Jackie' Fisher, championed the building of a revolutionary new type of battleship which would be superior in both firepower and speed to any existing battleship.

The Dreadnought Era

By March 1905, the design of the first ship – to be named *Dreadnought* – was complete. In May 1905, the Russian fleet was annihilated by the Japanese at the Battle of Tsushima. This action highlighted the importance of speed if an enemy was to be overtaken and brought into action, and demonstrated the hitting power of 12in guns, thus supporting Fisher's arguments.[1]

On 2 October 1905, the keel of the new ship was laid down in Portsmouth Dockyard. By then Fisher had been elevated to First Sea Lord and was able to ensure that construction of the new ship proceeded with extraordinary speed, so that *Dreadnought* was launched by King Edward VII just four months later, on 10 February 1906. All this effort was intended to show Germany how formidable the British shipbuilding industry was, and to give Britain an unassailable lead in battleship construction.

Dreadnought was 'completed' for preliminary trials on 3 October 1906, a year and a day after being laid down. This date was really for publicity purposes – final completion was actually two

months later. She was the largest, fastest and most heavily armed battleship in the world and, for the first time in a battleship, used steam turbines for propulsion. *Dreadnought* gave her name to the new breed of battleship and introduced the idea of an all big-gun battleship, rather than mounting guns with a combination of different calibres, an idea promulgated by the Italian naval architect Vittorio Cuniberti.[2] At a stroke, all of the world's pre-Dreadnought battleships were rendered obsolete, and the great naval arms race between Britain and Germany accelerated.

Larger, faster and more heavily armed, the Dreadnoughts were once again pushing the dockyard's capacity to its limits. Portsmouth's pre-eminence as a dockyard was unquestionable. The first of each new class of Dreadnoughts was built there – *Dreadnought* herself being followed by *Bellerophon* (1907), *St Vincent* (1908), *Neptune* (1909), *Orion* (1910), *King George V* (1911), *Iron Duke* (1912), *Queen Elizabeth* (1913) and *Royal Sovereign* (1915). Capital ships continually increased in size as the new classes of Dreadnoughts were introduced. The pre-Dreadnoughts of the King Edward VII class had a length of only 457ft and a beam of 78ft. *Dreadnought*

The battleship *Neptune* was built at Portsmouth and completed in 1911.

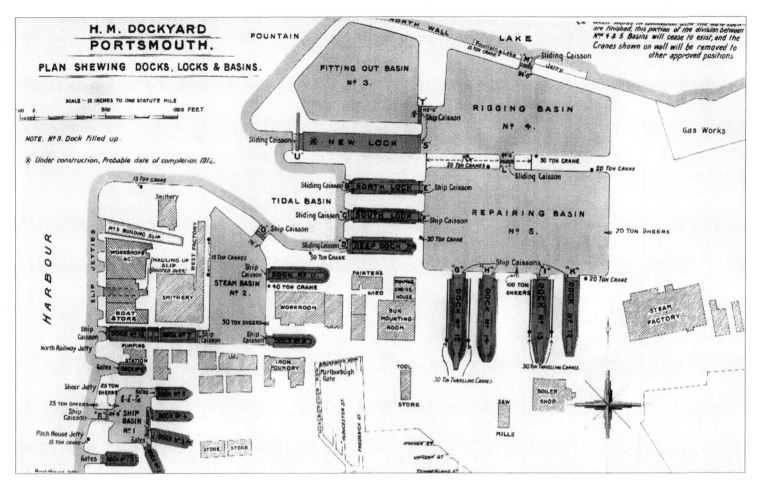

TA plan of the docks at Portsmouth Dockyard, 1909. The original format of the three basins (Nos 3–5) in the Great Extension is shown. (From *The Dock Book*, 1909, The Admiralty)

was 526ft in length, with a beam of 82ft, and her type had grown to 640ft with a beam of 90ft by the time of the Queen Elizabeth class. Even longer were the battlecruisers – the Renown class was 794ft in length, and *Hood* 860ft.[3]

At Portsmouth in 1906, only No. 15 Dock could accommodate *Dreadnought*, because of her beam. The South and North locks (later A and B) were too narrow for *Dreadnought*, so only the entrance to the rigging basin at Fountain Lake Jetty could be used to give access to No. 15 Dock, and this only at fourteen times a year, working at two tides a day: but this was reduced by the fact that docking was not carried out at night. When *Dreadnought* was at deep draught, she could only pass through this entrance on three days of the year.[4] The advent of the Dreadnoughts and the concurrent growth in the size of the Navy therefore led to the need for a new phase of capital investment in the dockyard.

The new plans to address this included combining the three separate new basins (Nos 3–5, being respectively the Repairing, Rigging and Fitting Out Basins) into one (known as No. 3), by dispensing with the coaling point and parade ground on the wall between Nos 3 and 4 Basins, and cutting back the wall between Nos 4 and 5 Basins, leaving a promontory extending out from the eastern wall of the basin, on which would be built a giant cantilever crane, capable of lifting 250 tons. This crane was electrically operated and would lift heavy machinery, guns, armour plate, etc, into and out of ships.[5] It was erected by Sir William Arrol & Co., of Dalmarnock, Glasgow, in 1912. A test load of 300 tons was applied every five years,[6] and this crane remained as one of the most prominent features of the dockyard until 1984.

Plans were made in 1906 for C Lock – a new dock and basin entrance – with provision for another lock (D Lock) to the north of it. The water over the sill entrance to the new lock was to be 46ft deep at high spring tide, compared to 34ft at No. 15 Dock, and the length would be 850ft (at the inner stops) or 921½ft at outer stops of caissons. A decision to proceed with D Lock (of the same size as C Lock) was made in 1910, and the North Pumping Station was built for the new locks. C Lock opened on 8 April 1913 with the battlecruiser *Princess Royal* entering, whilst D Lock followed in April 1914 with the docking of the battlecruiser *Queen Mary*. The capacity to dock the larger Dreadnoughts was also increased by enlarging two existing docks: No. 15 Dock was lengthened in 1907 to 613ft at the inner stops, whilst No. 14 Dock was lengthened in 1914 to 723ft.[7] To replace the Coaling Point, a coal depot ship that could hold 12,000 tons arrived in the yard in 1904.[8] Part of the Fitting Out Basin was also lost, the remaining part becoming known as The Pocket.

The mighty 250-ton crane erected in 1912 on the promontory in No. 3 Basin. This is a late view, probably *c.*1980, just a few years before the crane was dismantled. (Portsmouth Royal Dockyard Historical Trust)

Looking across C and D Locks towards Fountain Lake in 1914, shortly after the locks were completed. Each of the two locks is holding a pre-Dreadnought battleship. Note the newly completed No. 4 Pumping Station towards the bottom right of the picture.

No. 14 Dock was lengthened in 1914, when this photo was taken.

The Docks and Locks at Portsmouth

Dock/Lock	Opened	Lengthened	Closed	Length[i]	Width	Entrance	Depth
1	1801		21 Feb 1984	277ft 0in	92ft 6in	57ft 0in	27ft 2in
2	1802		1922	253ft 4in	89ft 2in		31ft 8in
3	1803	1859–60	8 Dec 1982	287ft 2in	81ft 9½in	67ft 5in	33ft 7in
4	1772	1859–60	7 Dec 1983	285ft 5in	73ft 2in	67ft 3in	33ft 0in
5	22 Jun 1698[ii]	1850	9 Nov 1983	259ft 7in	81ft 0in	55ft 1in	25ft 5in
6	1700[iii]	1908	3 Apr 1984	235ft 0in	82ft 10in	52ft 8in	25ft 2in
7	Aug 1849	1859[iv]	29 Feb 1984[v]	289ft 3in	89ft 9in	80ft 5in	37ft 2in
8	1853			240ft 4in	80ft 10in	70ft 1in	29ft 3in
9	24 Aug 1850		c.1898[vi]	295ft 3in	88ft 0in		26ft 1in
9	1875			440ft 7in	101ft 6in	81ft 9in	47ft 6in
10	1858	1859[vii]	16 Mar 1989[viii]	305ft 4in	95ft 1in	88ft 11in	37ft 1in
11	1 Jul 1865			427ft 0in	96ft 7in	69ft 9in	34ft 1in
12	1876	1903		500ft 9in	110 0in	80ft 0in	42ft 6in
13	Mar 1876	1905	2001[ix]	563ft 9in	110ft 0in	82ft 0in	42ft 8in
14	25 Jun1896	1914		723ft 0in	120ft 0in	100ft 0in	43ft 6in
15	1896	1907		612ft 11in	120ft 2in	93ft 11in	43ft 7in
A	1875			467ft 7in	100ft 4in	81ft 9in	47ft 8in
B	1875			467ft 7in	100ft 0in	82ft 0in	47ft 10in
C	8 Apr 1913			850ft 0in	119ft 0in	110ft 0in	55ft 8in
D	Apr 1914			850ft 0in	119ft 0in	110ft 0in	55ft 8in

Notes:

i   Length of dock is with caisson on inner stop.

ii Originally the Great Stone Dock, with dimensions 190ft x 66ft and with a 51ft entrance. Rebuilt in 1769.

iii Originally the Lower Dry Dock, completely rebuilt in 1737.

iv Converted to a continuation dock combining Nos 7 and 10 Docks.

v   Filled in 1989.

vi Filled in by 1900.

vii Converted to a continuation dock combining Nos 7 and 10 Docks.

viii Filled in 1989.

ix Filled in 2002 for VT Group.

(Sources: Patterson, 1989, p.57; Coad, 2013, pp.176–78.)

In 1907, a new Factory was completed in a position near the Unicorn Gate at the east end of the yard. It was originally constructed with a view to the repair of steam reciprocating engines and their auxiliary machinery, but was altered to deal with large repairs to turbines, and consisted of five bays containing the Torpedo Tube Shop, Erecting Shop and Fitting Shop. The machine tooling included a large quadruple-geared motor-driven lathe which could machine the rotors of large turbines, taking jobs of up to 55ft between centres; a large boring machine could bore turbine casings up to 17ft in diameter, taking a job 45ft between

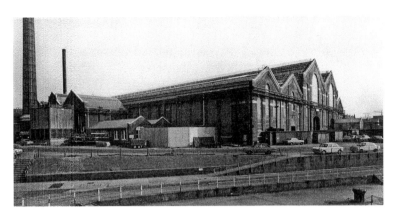

The new factory, known as the MED Factory, was completed in 1907. (Portsmouth Royal Dockyard Historical Trust)

centres; and a wall planer was capable of machining a job 20ft wide and 30ft high. Overhead cranes of 80 tons capacity were provided to lift turbine casings and other items into and out of the machines. Known at first as the New Steam Factory, and later as the MED Factory, it measured 580ft by 280ft in plan, and many of the machines in the new facility had been transferred from the old Steam Factory, which had been appropriated to form an extension to the machinery shops of the Ship Construction Department.[9] The old Steam Factory was renamed the West Factory.

To the south-east of C Lock, the elegant two-storey North Pumping Station (No. 4) was constructed in 1913, together with its single-storey boiler house, both built of red brick with Portland stone arches, lintels, copings and dressings. The steam engine house was equipped with two main dock pumps capable of pumping 15,000 tons of water per hour.[10] A new 150-ton lift electric overhead travelling crane was built by Messrs Vaughan & Sons Ltd and erected at the Guns Store Ground in the north-east corner of the yard in 1914.[11] An Admiralty Floating Dock, later known as AFD 5, with a lift capacity of 32,000 tons, was completed by Cammell Laird in 1912 and towed to Portsmouth. It was equipped with its own workshops, electricity generators, both electric and steam cranes and eight sets of pumps able to move 48,750 tons of water in four hours.[12] At the time it was the world's largest floating dock and, together with the two new locks and the two enlarged docks, it gave Portsmouth the capacity to dock five Dreadnoughts simultaneously. A new 150-ton lift floating crane was also provided to work alongside the dock, and a jetty was built in the dockyard at the eastern end of Fountain Lake with a railway link so that repair materials could be brought alongside the floating dock.[13] This dock, with the floating crane docked inside it, was towed to Invergordon in September 1914, to provide a facility capable of taking Dreadnoughts in the north, pending completion of Rosyth dockyard.[14]

Fire in the Semaphore Tower

At 7.30 p.m. on the evening of Saturday, 20 December 1913, a fishmonger's delivery boy, on leaving the new battlecruiser *Queen Mary* moored alongside South Railway Jetty, noticed a flickering light at one of the Sail Loft windows. He brought this to the attention of the ship's quartermaster, who realised the situation and raised the alarm. Up above, in the wooden Semaphore Tower, three signalmen were on duty and apparently unaware of the conflagration below. Within minutes, men of the Metropolitan Dockyard Police were on the scene, working hydrants within the building, but the fire was so fierce that they proved useless. As the smoke was funnelled up the tower, two of the signalmen (who were trying to descend) were overcome by the fumes and perished at the scene. The third signalman managed to escape. By 9.30 p.m., the tower had collapsed into the main building.

More than 1,000 men, including the Metropolitan Dockyard Fire Brigade, sailors and firemen from the Royal Marine Barracks at Eastney and Forton, attended the blaze. The fire engine from the Royal Clarence Victualling Yard hurried across Portsmouth Harbour's floating bridge. Many soldiers and civilians ran to help, whilst fire parties from fifteen ships in the harbour also attended. But it was to no avail, as the inferno took hold and the men who fought the blaze were driven back by the flames. The fire had started in the Sail Loft in the south wing of the building and spread to the adjacent Semaphore Tower (in the middle of the building, with an arch below) and the Rigging House in the north wing. The building was 400ft long by 50ft wide, and stored inside was stock normally found in a sail loft or rigging room, including sailcloth

The old Semaphore Tower (1833) which was destroyed by fire in 1913. (Portsmouth Royal Dockyard Historical Trust)

and tarred rope. All this simply fuelled the fire, and 200 sewing machines in the Sail Loft were destroyed, along with the roof and upper floor.

The *Queen Mary*, enveloped in smoke, with sparks and debris falling on her deck, was moved up the harbour by tugs. With wooden decking and the ship's boats stowed on the upper deck, there was an immediate danger to her. It wasn't until 3 a.m. that the fire was brought under control, and hoses were played on the embers for the remainder of the night. At daybreak, the gaunt remains of the lower square tower, below the collapsed signal section, stood firm but in danger of collapse and were ready to topple at any moment. It was said at the time to be the most disastrous fire in the dockyard within living memory. Built in 1833, the main tower structure was one of the most prominent features

on the waterfront and was well known to everyone who passed by and underneath it. It had last been used to send semaphore messages to the Admiralty in Whitehall on 31 December 1847, and had was thereafter used as a signal station.[15]

Fisher's Reforms

The total workforce in all of the royal dockyards had grown from 19,000 to 33,700 (an increase of 77 per cent) between 1895 and 1905, but by this time the extent of shirking in the yards was officially reported to have reached scandalous proportions.[16] In 1904, the House of Commons was told that, many industrious men notwithstanding, the dockyards were a refuge for men who 'did their work in a malingering way, and who did not do a great deal of work'. The growth in size of the workforce had clearly outstripped the ability of existing systems to cope.[17] There followed retrenchment in the dockyards at the instigation of the First Lord, the Earl of Selborne, and Admiral Sir John Fisher, the new First Sea Lord, who chaired the Naval Establishments Enquiry Committee, which was influenced by practice in commercial yards and new American ideas on decentralised management. One of the committee described the situation he had seen at Devonport:

> I found it a huge bureaucracy, in consequence almost nerveless, real energy, initiative or true responsibility had been obliterated; and the object of everyone was to save their face by compliance with the Regulations ... each [department] struggling for its own, the efficiency of the yard as a whole was lost sight of.[18]

A more decentralised structure was introduced, giving 'full authority' to the chief constructor and chief engineer within their departments, and changing their titles to Manager, Constructive Department and Manager, Engineering Department. Both still reported to the Admiral Superintendent, who was supposed to take a more arm's-length 'owner' role. The role of the Director of Dockyards, which had been created in 1872, and by this time consisted mainly of office work at the Admiralty, was changed to allow closer personal technical supervision in the yards, and its title was changed to Director of Dockyards and Dockyard Work (it was

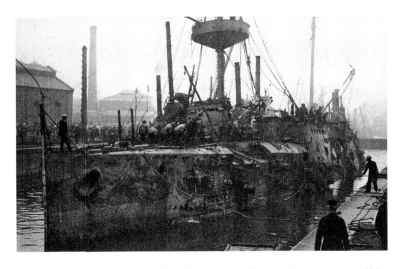

Docking the cruiser *Gladiator* with collision damage. During a late snowstorm off the Isle of Wight on 25 April 1908, *Gladiator* was heading into port when she struck the outbound American steamer SS *Saint Paul* off Point Hurst. *Gladiator* settled on her starboard side in shallow water close to Fort Victoria. Salvage work began almost at once, but it took over five months to right the ship, refloat it and tow her back to Portsmouth. The damage was extensive and as the ship's design was obsolescent, she was scrapped rather than repaired. She had been built at Portsmouth and completed in 1899. (R. Silk)

further strengthened in 1912, with more direct responsibility, following the abolition of the Controller's Department).[19]

In 1904–05 and 1905–06, cuts in dockyard budgets led to a total of 7,000 job losses:[20] at Portsmouth, 1,500 men were discharged in 1905.[21] Redundancy pay was generous by the standards of the day: the *Morning Telegraph*, though, was critical, reporting:

> [E]very man having served seven years or more, on discharge gets a week['s] pay for each year, a minimum of £7. But many men get £1.15s a week, and some of them have served 30 and 40 years which means that they will get between £50 and £70 for doing no work. Do you call this economy?[22]

The reduction in dockyard jobs was, partly, a consequence of a reorganisation of the fleet as part of Fisher's reforms: numerous ineffective ships were struck off the Navy List (Fisher called them 'ships that could neither fight nor run away'), and the system for repairing ships was altered, as Fisher explained:

> The old plan was for ships to be drawn to the verge of inefficiency at sea, to be then allowed to vegetate for an indefinite period of time in the Dockyard Reserve, and finally (at some time that suited Dockyard arrangements) to be pulled to pieces and then put together again"[23]

This approach was replaced by a system of planned maintenance so that, 'ships are to be kept as a going concern throughout their lives'. Each ship was to have a thirty to forty-day annual refit, and every fourth year there would be a 'moderate refit' or possibly (and less frequently) a 'large' refit if extensive modernisation or reconstruction was necessary. Ships in refit would retain a nucleus crew, rather than being paid off completely, and this crew would carry out care and maintenance tasks and give assistance to dockyard employees. The sea-going service would dictate priorities and the extent of defect lists, rather than, as before, the dockyards. Furthermore, all ships above destroyer size now had their own workshops on board which could carry out many repairs, whilst the smaller vessels had repair ships (a recent innovation) which could undertake such work. Periodic surveys to ships, which involved stripping and practically rebuilding, were abolished. Refits of the ships in a squadron would be staggered, rather than the previous practice of all the ships entering the dockyards at the same time.

One report considered that the achievement of the reductions in manpower was 'a wonder to this day'. A meeting of Commanders-in-Chief Home Ports and Admirals Superintendent received a memorandum from Fisher which noted that:

> To get rid of a Dockyard workman involves agitation in every direction – in Parliament, at the Treasury, and locally, and even Bishops throw themselves into the fray, like the Bishop of Winchester at Portsmouth, instead of looking after his own disorganised and mutinous Established Church.[24]

Following the reductions, the dockyards and their supporters lobbied the Admiralty, saying that the workforce was insufficient and that arrears of work were building up. Fisher dismissed these claims, saying that temporary recruitment to deal with peaks in workload

The viaduct connecting the South Railway Jetty to the main line at the north end of the Harbour station. On the left, a locomotive named *His Majesty* heads a royal train which is about to depart from the jetty, whilst sailors aboard a capital ship look on, in 1910.

was not practicable because dockyards could not discharge men with as little trouble as private employers. Over time there would be peaks and troughs, and the work would average itself out.[25]

Other, more minor, reforms were introduced to address inefficiencies. Stockholdings in the dockyard stores totalled £22 million, including enormous stocks of redundant items no longer used by the Navy and things that could be purchased at short notice such as chairs (10,000 of these were in stock). On the other hand, many condemned stores were found to have considerable life left in them, but ships were reluctant to take stores that were not perfectly new. Efficiencies were sought in these areas; to reduce the cost of some items, trade patterns were introduced where practicable, avoiding the cost of special manufacture to meet naval specifications (that were in fact often out of date).[26]

Many suggestions were made to improve workflow and methods in the dockyards, often requiring investment in new machinery and handling equipment.[27] These ideas were consistent with Fisher's idea of rapid shipbuilding, which was intended to result in ships spending a shorter time on the building slips and in the fitting-out docks and basins, giving fuller use of dockyard facilities and earlier utilisation of the capital spent in new construction. He sought to achieve this 'by concentrating all the workmen then leisurely building several different ships, and put them all like a hive of bees on to one ship, and extended piece-work to the utmost limit that was conceivable'.[28]

After the cuts of 1904–05 and 1905–06, the dockyards once more entered into a period of growth as a rearmament programme gathered momentum. However, unemployment in Portsmouth remained high and in October 1908 the prime minister announced that 960 men would be taken on in the dockyard as a temporary measure to help alleviate the situation. A further 286 were specially entered to expedite the completion of work on HMS *Terrible*. By late March 1909, these men were on notice of discharge, though some were retained.[29] By 1913–14, total expenditure was £1,135,000 at Portsmouth Dockyard, a 45.6 per cent increase when compared with 1905–06,[30] or 36.5 per cent when adjusted for inflation.[31]

New Construction

Between 1895 and 1906, nine pre-Dreadnought battleships had been built at Chatham, eight at Portsmouth and seven at Devonport, whilst Pembroke had built just one; 62.5 per cent of these pre-Dreadnoughts were built in royal yards, compared with only 36 per cent of the subsequent Dreadnoughts and battlecruisers (the remainder being built by private yards). Between 1906 and 1916, Portsmouth built nine Dreadnoughts, including the lead ship for every class of battleship. Devonport built six Dreadnoughts, plus two battlecruisers (whose extra length could be accommodated there). Pembroke and Chatham built none.[32] In 1913–14, 31 per cent of total budgeted expenditure on new construction (including all costs) was for dockyard-built ships. However the dockyard-built

The Nasmyth steam hammer in the Smithery. (Portsmouth Royal Dockyard Historical Trust)

ships included much contract work on guns, gun mountings, turbines, boilers and armour plate – all of which were manufactured by private firms. This contract work comprised 62 per cent of the ships' total cost.[33]

At this time there was concern about the late delivery of new ships: the Admiralty specified completion of battleships within two years of the keel being laid, but in 1912–13, only one of the seven capital ships completed met this target. This was *King George V*, which was built by Portsmouth in twenty-two months. One factor was that dockyards received the plans at an earlier date than private yards, allowing more work to be done in preparing material before laying the keel, but nevertheless Portsmouth's performance was considered to be very good. Employee discipline and regularity of attendance was said to be better in dockyards than in private yards.[34]

The causes for the delays were said to include industrial disputes and shortages of skilled men. The numbers of apprentices recruited were being limited by the trades unions and were insufficient to meet natural wastage of 'journeymen'. Many trained men had emigrated to the colonies, especially Canada, during the slump of the mid-Edwardian period and had often taken up employment in agriculture.[35] It was pointed out that the size and complexity of capital ships had increased, making it more difficult to meet the delivery target. A pre-Dreadnought had required 5.27 million man-hours to build, whilst *Dreadnought* required 5.49 million man-hours and *Iron Duke* 7.2 million man-hours. In the case of *Iron Duke*, this would occupy 2,800 men for two years. However, the introduction of labour-saving machinery and the fact that turbines required far fewer man-hours than reciprocating engines of the same horse-power output had helped limit the increase in total man-hours required.[36] No general improvement in completion times was seen in subsequent years: only one more battleship was completed in less than twenty-four months – this being *Iron Duke* at Portsmouth.[37]

Expenditure on new construction at Portsmouth increased from £367,000 (excluding armour and other contracted work) in 1905–06 to £595,000 in 1913–14, an increase of 62 per cent (52 per cent in real terms). Simultaneously, expenditure on repairs and reconstruction of ships at Portsmouth increased by 53 per cent (44 per cent in real terms) to £664,000.[38]

Preparations for the launch of the battleship *St Vincent*, 10 September 1908. A Royal Marines band is positioned below the bow, and to the right covered steps leading up to the launching platform can be seen, guarded by a policeman. (Portsmouth Royal Dockyard Historical Trust)

The keel of *Queen Elizabeth* was laid down on Trafalgar Day, 21 October 1912. She was the first of a new class of five super-Dreadnoughts that would be bigger, faster, more powerfully armed and more heavily armoured than any of her predecessors. And, for the first time in a British battleship, she would be oil-fired rather than coal-fired. Whilst the Admiralty was nervous about being reliant on imported oil rather than home-mined coal, Winston Churchill, as First Lord of the Admiralty, persuaded the government to invest in the Persian oilfields to secure supply. To achieve a speed of 24 knots, some 3 knots faster than the preceding Iron Duke class, required doubling the power output of the steam turbines and employing twenty-four boilers, six more than *Iron Duke*. The boilers were arranged in four adjacent compartments, aft of which were four engine rooms placed abreast.[39]

In July 1914, just before the outbreak of the First World War, there was a review of the fleet at Spithead where the newly completed *Iron Duke* joined fifty-four other battleships and four battlecruisers, plus a host of cruisers, destroyers, submarines and torpedo boats, in a forbidding spectacle of sea power that eclipsed all previous fleet reviews at Spithead. When the ships dispersed, the First Fleet sailed north to its wartime anchorage at Scapa Flow.

> Do not believe what people tell you of the ugliness of steam, nor join those who lament the old sailing days. There is one beauty of the sun and another of the moon, and we must be thankful for both.
>
> Rudyard Kipling (1865–1936), *A Fleet in Being*

Design

In 1905, a high-level design committee, chaired by Admiral Sir John Fisher, was set up to advise on the design of the revolutionary battleship proposed by him. It contained senior naval constructors (from both the Admiralty and private shipyards) to advise on constructive and engineering aspects, and senior naval officers to advise on the operational requirements for the ship, including the tactical and fighting aspects. One of the most influential members was Sir Philip Watts, Director of Naval Construction, who had served his apprenticeship in Portsmouth Dockyard. The committee met for the first time on 3 January 1905 and agreed its report seven weeks later, on 22 February, an incredibly short time for the number of decisions made. At the first meeting, Fisher defined the governing conditions of the design as guns, to be 12in, and speed, to be 21 knots, saying, 'Both theory and the actual experience of war dictate a uniform armament of the largest guns, combined with a speed exceeding that of the enemy, so as to be able to force an action.' The committee's report made it clear that the reason for an all big gun armament was concentration of force, to bring an overwhelming weight of heavy guns on a part of the enemy's battle line. After considering various alternatives for the number and layout of the twin 12in gun turrets, a layout of five twin 12in turrets was decided upon. The merits of both reciprocating and turbine engines were considered before settling on the latter. Turbines had been trialled in destroyers and cruisers, and would give a total weight saving of around 1,000 tons and produce less vibration.[1] Driven by Fisher's enthusiasm for innovation, the ship's turbines, hull design, firepower, range finding and communication systems were to make her superior to all earlier battleships. Fisher was, however, blinkered by his longstanding feud with Admiral Lord Charles Beresford, preventing the installation of the innovative Argo fire-control system devised by Arthur Pollen, Beresford's protégé.[2]

Work by naval constructors in Whitehall on the detailed design then began. A hull form was designed and tested in model form, at the Admiralty's experimental ship tank (which had been installed in 1886 at Haslar, Gosport), and various modifications were made, with seven models being used in all. At the time there was a measles epidemic and the men carried out these tests in quarantine. For the detailed plans, a team of two assistant constructors and five draughtsmen was initially deployed, and this built up to peak at three assistant constructors (who each had their own briefs) and thirteen draughtsmen.[3] The hull structure and fittings were designed to continue the steady reduction in weight and other detailed improvements which had been made in recent years. One major change was the introduction of longitudinal bulkheads to

The Mould Loft Floor. (Portsmouth Naval Base Property Trust)

protect the magazines from underwater explosions. It was decided that the officer accommodation should be forward, near the bridge, instead of in the conventional space aft.[4] It was intended to build the ship in record time:

> In order to facilitate this the hull structure was designed so as to secure the utmost simplicity in every way while retaining the exceptional rigidity required in the decks and bulkheads to successfully withstand the shock of firing such a large number of heavy guns. The variety of steel sections used was reduced to a minimum; and a great effort was also made to adopt plates, of the several thicknesses required, in standard sizes, so as to avoid the usual labour and waste of time in cutting and sorting plates. Considerable quantities of plates, etc, were ordered to standard sizes and by this means an ample supply of materials was maintained well in advance of the work on the ship.[5]

Bending frames on the slab in the Smithery. (Portsmouth Naval Base Property Trust)

Construction Programme

Manufacturing work in the dockyard started in early May 1905, and in the five months before the official laying down date – Monday 2 October 1905 – £29,078 had been spent on materials and £12,217 on labour, about 6,000 man weeks of work, at forty-eight hours a week, with 1,100 men being engaged by the laying down date. The job of the Chief Constructor at Portsmouth, Thomas Mitchell, was retitled Manager, Constructive Department, under Fisher's dockyard reforms, and he was made responsible for ordering all the materials for the ship. At the time of the launching, he was made a Commander of the Victorian Order, and upon his retirement in 1907 he was knighted[6] for his work on *Dreadnought*, probably the only knighthood bestowed so low down the seniority list of the Royal Corps of Naval Constructors, and even more notable in that he had been promoted from the foreman grade. Mr E.J. Maginess was the constructor in direct charge of building, whilst Mr J.R. Bond planned the work in the mould loft.[7]

The frames of the ship were prefabricated in the dockyard, allowing them to be erected immediately after the keel was laid on No. 5 Slip – with bulkhead erection starting just three days after the keel laying. By 7 October (day six) the majority of the middle-deck beams and the first of the bulkheads, were in place.

The number of men employed on the work increased from 1,100 to 3,000. Instead of the usual forty-eight hours, dockyard workers worked compulsory overtime with a six-day week of sixty-nine hours, which was 6 a.m. to 6 p.m. each day with only half an hour for lunch, using searchlights and lucal lamps (pressurised lamps burning a spray of paraffin and petroleum that produced a large 6ft flame similar to a blow lamp) to illuminate the work later in the day instead of finishing at 3 p.m. or 4 p.m., and some men even worked on Christmas Day and New Year's Day. The long hours were not popular with the men: double shift working was considered, but there was insufficient manpower to make this possible.[8]

By day thirteen, most of the middle deck protective plating (1¾in flat, 2¾in slope) was in place, and by day twenty (Trafalgar Day, 21 October) the hull plating was well advanced and the stem was in place. By day fifty-five, the beams of the upper deck were in place. By day eighty-three, the upper deck plates were in place and work was starting on the forecastle. On day 125, six days before the launch, work on the hull was finished: the ship was launched by King Edward VII on 10 February 1906, just over four months after laying down, a remarkable achievement.[9]

The Construction Process[10]

Copies of the plans were sent from Whitehall to the dockyard by couriers. They were used by the Mould Loft to create light wooden templates showing the exact shape of every shell plate and frame. The Mould Loft at Portsmouth had been built in 1893 (and enlarged in 1908), measured 150ft x 100ft and was a large open plan space with no inner supporting columns. The floor was of yellow pine, known as the scrieve boards, on which the full-sized shapes of the frames and plates were drawn by a system of dots and chalk lines. The lines were then cut into the scrieve boards using scrieve knives. A light batten secured by pins was placed on the scrieve board profile and formed to shape. The loftsmen were shipwrights, assisted by a small number of apprentices. A model of the hull was made to 1:48 scale, and on it the positions of the shell plates were worked out to minimise plates curved in two dimensions and reduce scrap from non-rectangular plates.

The steel used for plate and sections (stiffeners) was of four grades: mild steel; nickel-steel for the decks and protective bulkheads; high-tension steel for the upper deck; and armoured plate for the armour. It was supplied by steel mills in Sheffield and Glasgow. The rolled plate was sent to the dockyard's plate shop, where it was pickled in acid tanks to remove millscale and passed through plate straightening rolls to remove any slight waviness produced when the plates cooled after rolling at the steel mill. Where necessary, the plates were bent using hydraulic bending machines to match the templates from the Mould Loft. The rivet holes were then drilled or punched, and flanges and bevels were machined, before the plates were cut to their final dimensions by shears. Armour plate was bent and machined by the armour-makers rather than the dockyard. Frames (side stiffeners) and bars were bent on steel slabs in the dockyard smithery using the sett (thin steel templates) for the correct shape. The slab was perforated with hundreds of large holes 1½in in diameter and 6in apart, which took the steel plugs around which the pre-heated frame was bent using levers, hammers and iron bars to force the appropriate fit. It was held down by dogs located in the slab's holes until it had cooled to the correct shape. The formed plates and stiffeners were now ready to go to the building slip. No. 5 Slip at Portsmouth, where *Dreadnought* was built, had a series of

Sorting and marking plates in the Erecting Shop. (Portsmouth Naval Base Property Trust)

pole derricks supported by guys at intervals on each side of the berth along most of the length of the hull. These derricks lifted the plates, sections and castings into place.

Blocks on which the keel would sit were placed on the slip and carefully aligned. The first 20ft flat keel plate was slung and landed in position, which was accompanied by the keel-laying ceremony. Once the full keel was laid, work proceeded on the construction of the double bottom with its complex network of frames, longitudinals and beams. The prefabricated frames were then erected, together with longitudinal stringers, riveted to floor plates and secured to the keel. They were bracketed to the transverse beams. Vertical and longitudinal bulkheads were also erected on the double bottom to form compartments, usually watertight, such as machinery spaces and magazines. Fore and aft girders stretched between the tops of the bulkheads to support the deck plates. This continued upwards, deck by deck, until the weather deck was reached. Large openings were left in the upper decks to allow the machinery, barbettes and gun mountings to be installed after launching.

The ship under construction on the slip, seen just five days after the keel laying.

the other side of the plate and knocked the rivet down with alternate hammer blows. Millions of rivets were needed on the ship and the work was extremely arduous, with workers often exposed to the full force of the winter weather. It was also frequently dangerous, with the gang working high up on the puddocks on insecure deck plating. To make the rivet joint watertight, it was caulked: the edges of the metal plates were chinzed using a hammer and chisel to compress the plate's edge, closing any gaps. It was then injected with red lead paste. Lap and butt joints between plates also had to be caulked in a similar fashion.

The responsibilities of the shipwrights during *Dreadnought*'s construction included: ensuring that the ship's lines were laid off correctly on the mould loft floor; setting up the slipways; laying the keel; setting up the frames and moulding plates; installing fittings which pierce the skin of the ship, e.g. port-holes, hawse pipes, hatches, propeller shaft openings; all wood work apart from cabins, including the decking; seating for all machinery; and launching and dry docking the ship.

The Launching Process[11]

Before launching, the hull was painted. Wooden cradles, called poppets, were made to fit the shape of the hull near the bow and at the stern, to transfer the weight out onto the launchways (or ways). The two fixed launchways, which were about 5–6ft wide, were manoeuvred under the hull, one on each side, each about 15–20ft from the centreline. Sliding ways and the poppets were temporarily attached to the hull, and a layer of grease and tallow was liberally applied to the bearing surfaces of them and the launchways. Some 500 shipwrights hammered long wooden wedges along the ship's entire length to raise the ship slightly and transfer the weight from the building blocks to the cradles and launchways. The hammering was commenced from the stern and was co-ordinated by the chimes of a bell. Just before launch there were only wooden baulks, known as dogshores, preventing the ship from sliding down the launchways. The king then named and blessed the ship, before cutting the cord which released the half-ton weights that knocked away the last of the dogshores. The slip had a declivity of about one-in-nineteen, allowing it to slide down it by gravity.

The ship included a number of castings and forgings for complex-shaped objects, and these were either made in the dockyard's foundries and smithery or by specialist private contractors. Examples of castings were stern frames to support the rudder, the stem piece which also formed the ram bow, and shaft brackets to support the propeller shafts. Where heavy castings and forgings, or items like bulkhead armour plates, had to be lowered into the ship, then temporary sheer legs were used for lifting because the slipway derricks had insufficient lift capacity.

Shell plates were bolted to the frames (as a temporary measure to hold them in place), faired, riveted and caulked. Riveting and caulking was undertaken by skilled labourers, rather than shipwrights. They typically worked in a gang of four: two riveters – a left-handed and right-handed striker – a holder-up and a heater, who was often a boy. The heater heated the rivets in a small coke portable furnace until they were white hot, and then tossed them to the holder-up, who pushed them into the hole and held them there with a long hammer known as a dolly. In shell riveting, the two riveters were on

The Launching Ceremony[12]

Elaborate plans were made for the launch ceremony, which was to be on a Saturday so that it could be a festive occasion viewed by a large crowd. The town, harbour and dockyard were to be decorated in flags and bunting, with arches erected to welcome the king, and naval and marine bands would be positioned on the route taken by the king, providing suitable music for the occasion. However, on 29 January, twelve days before the launch, Queen Alexandra's father, the King of Denmark, died and the court was still in mourning on the day of the launch. The lavish displays of colour and music were abandoned, except on the launching platform itself; the queen did not accompany her husband as had been planned, and the ceremony became a more sober affair.

Just over 1,000 guests obtained tickets for the launching platform, and spectators came long distances to see the ceremony. A heavily laden special train brought a contingent from London, and at 10.30 a.m. the dockyard gates were opened in order to allow the wives and children of employees, as well as the general public, to see the launch, whilst the *Seahorse* and the *Magnet* brought naval cadets from Osborne. The crowd around the slip was enormous, *The Times* saying the number was so vast it could not be estimated. A south-west gale had raged during the night, but luckily it had subsided somewhat by morning and the sun shone fitfully through scudding clouds. Marines and seamen lined the king's route from the royal yacht at the South Railway Jetty, as the royal train steamed to a specially erected station at the slip. The king, in the uniform of Admiral of the Fleet, was accompanied on the train by the First Lord of the Admiralty and the Commander-in-Chief, Portsmouth, and was received at the slip by the Board of Admiralty. There were gathered all the dignitaries and senior naval officers, the latter in full dress with cloaks, mourning bands also being worn. Amongst the many distinguished guests in the king's enclosure were Admiral Sir John Fisher, the First Sea Lord; Sir Philip Watts, Director of Naval Construction; Mr J.B. Marshall, Director of Dockyards; and Fisher's *bête noire*, the Bishop of Winchester. There were ten foreign naval attachés present, including Rear admiral Carl Coerper, representing the German Empire, who was no doubt taking a great interest in proceedings. The king mounted the rostrum and a short religious service followed.

On 3 February 1906, 125 days after keel laying, the ship is ready for launching, with the bow cradle in position. In the foreground, stands for spectators are being erected. (Portsmouth Naval Base Property Trust)

On day 125, shipwrights are setting the ship up, driving wedges in to raise the ship and transfer the weight from the building blocks to the cradles and ways. (Portsmouth Naval Base Property Trust)

The prow of the *Dreadnought* was decorated with a garland of flowers woven out of red geraniums, white tulips and smilax. This concealed a bottle of colonial wine with which the christening was to be performed. A handsome casket holding the chisel and mallet for the king's launching of the vessel was made out of Danzig oak taken from the pillars of the *Victory*; it was beautifully carved, as were the mallet and chisel, all designed and made in the dockyard. The managers of the constructive and engineering departments were presented to the king by Rear admiral Barry, Admiral Superintendent of the dockyard, and then the last blocks on which the ship rested were knocked away by shipwrights. The king took the flower-garlanded bottle of wine and dashed it against the ship, where it failed to break, but a second attempt was more successful. The king took the chisel and mallet and cut the cord, releasing heavy weights and knocking away the dog-shores holding back the ship. There was a blast from a bugle and the great ship began to move, sliding slowly down the launchways into the waters, accompanied by a band playing 'God Save the King', the only occasion on which a band was heard during the king's visit. Admiral Barry called for three cheers for His Majesty, which was responded to with enthusiasm. The *Dreadnought* by now was fully afloat, flying the Royal Standard, the Admiralty flag, the Union Jack and the White Ensign from her masts. Dockyard tugs took charge of the ship and moved her to her berth. The royal train conveyed the king back to the *Victoria and Albert* for an official luncheon before the yacht sailed for Cowes.

Fitting Out

Fitting out then proceeded: only No. 15 Dock could accommodate *Dreadnought*, because of her beam, and this was the dock she entered during fitting out. New steam-operated sheerlegs capable of lifting 100 tons had been provided in 1901[13] and were used to ship *Dreadnought*'s machinery and armament.[14] When she was not required to be in dry dock, she was alongside in the basin.

Inside the hull, decks were divided into cabins, messes, wardrooms, sick bays, officers' quarters, storage rooms, washrooms, galleys, magazines, workshops and so on, and ladderways were installed. The superstructure, including the funnels, masts, conning tower and compass platform, were all erected. A network of ventilation trunking was installed.

The 12in guns and the hydraulic machinery had been ordered in January 1905, the main propelling machinery on 24 June and the armour and principal ship castings in August 1905. The two sets of direct drive turbines were built by Parsons Marine Steam Turbine Co. Ltd, on the Tyne, the eighteen boilers by Babcock and Wilcox, Renfrew, and the armour plate was supplied by William Beardmore & Co. Ltd, Dalmuir. The 12in guns were made by Vickers, Sons and Maxim Ltd, Sheffield, and two gun mountings by Vickers, Sons and Maxim Ltd, Barrow, with the other three by Sir W.G. Armstrong, Whitworth & Co. Ltd, Elswick. All these items, as well as a host of others such as the torpedo tubes, secondary 12-pdr armament, director, rangefinders, auxiliary machinery, electrical equipment and wiring, pipework, hydraulics, steering gear, anchors, winches, capstans, pumps, boats and boat hoists, ammunition hoists, searchlights, ropes and cables, and internal fittings were installed during the fitting out. Gangs of men of different trades worked in a whole range of work sites around the ship. The amount of electrical equipment in battleships was increasing rapidly at this time. *Dreadnought* had, *inter alia*, five electric lifts, eight electric coaling winches, electric pumps, ventilation fans, telephones and lighting.

Dreadnought fitting out in No. 3 Basin, with sheer legs visible on the jetty behind.

Four Siemens dynamos (two driven by Brotherhood steam piston engines and two by Mirrlees diesel engines) with a total power of 410kW supplied the 100-volt and 15-volt DC circuits.[15]

The ship undertook basin trials on 17 September 1906 and was docked for a hull inspection from 24–28 September. On 1 October, she raised steam for 13 knots and proceeded out of the harbour for Spithead for two days of trials. On the following day, 3 October, a year and a day after being laid down, she sailed for Devonport for a full range of steam trials. It was on this date that the claim that she had been built in a year and a day was based. On 9 October, she averaged a speed of 20.05 knots during her eight-hour full-power contractor's trials off Polperro, Cornwall, and 21.6 knots on the measured mile. She returned to Portsmouth for gun and torpedo trials, then entered the dockyard on 19 October for final fitting out. On 1–2 December, she carried out a twenty-four-hour acceptance trial and then returned to No. 5 Basin. Between 10–17 December, she was in No. 15 Dock, but was commissioned into the fleet on 11 December 1906, her official completion date.[16]

The total cost of the ship was £1,785,683, which was made up as follows – Hull, fittings and equipment: labour £198,513, materials £535,914, contract work (armour) £110,357; Propelling and other machinery: dockyard work £8,750, contract work £310,835; Gun mountings, torpedo tubes, etc: dockyard work £7,940, contract work £382,205; Incidental charges £117,969; Guns £113,200.[17] Incidental costs covered various Admiralty and dockyard overheads. The value of contract work (which included the guns) was 51.3 per cent of the total cost.

On 5 January 1907, she left Portsmouth for an experimental cruise to Spain, Gibraltar and Trinidad, and was based at Port of Spain from 5 February to 18 March.[18] At a stroke, all of the world's pre-Dreadnought battleships were made obsolete, and the great naval arms race between Britain and Germany accelerated. There were, however, critics: Britain had the largest fleet of pre-Dreadnoughts but now had to rebuild the fleet from scratch, to the consternation of the incoming Liberal government. Sir William White, the former Director of Naval Construction, suggested that Britain was 'putting all her naval eggs in one or two vast, costly, majestic, but vulnerable baskets', and favoured larger numbers of smaller battleships.

Nevertheless, Fisher was undaunted: between February 1906 and February 1907, six more Dreadnoughts were laid down in

Britain – three of the Bellerophon class, which were virtually repeats of Dreadnought, and three of the more lightly armoured Invincible-class battlecruisers. In the next twelve months, the three St Vincent-class battleships were laid down, and the British press and public threw themselves behind the Dreadnought building programme. Meanwhile, Germany laid down her first four Dreadnoughts, of the Nassau class, in the summer of 1907.

Dreadnoughts built at Portsmouth Dockyard

Name	Laid Down	Launched	Completed	Fate
Dreadnought	2 Oct 05	10 Feb 06	Dec 06	Sold for BU 1921
Bellerophon	3 Dec 06	27 Jul 07	Feb 09	Sold for BU 1921
St Vincent	30 Dec 07	10 Sep 08	May 09	Sold for BU 1921
Neptune	19 Jan 09	30 Sep 09	Jan 11	Sold for BU 1922
Orion	29 Nov 09	20 Aug 10	Jan 12	Sold for BU 1922
King George V	16 Jan 11	9 Oct 11	Nov 12	Sold for BU 1926
Iron Duke	12 Jan 12	12 Oct 12	Mar 14	Sold for BU 1946
Queen Elizabeth	21 Oct 12	16 Oct 13	Jan 15	Sold for BU 1948
Royal Sovereign	15 Jan 14	29 Apr 15	May 16	Sold for BU 1949[19]

In 1908, the clamour for more ships reached fever pitch in Britain, with a panic deriving from evidence of increases to German Dreadnought-building capacity, and rumours that Germany was laying down Dreadnoughts in secret which would allow it to reach parity with Britain by 1912. A Tory MP coined the phrase 'We want eight and we won't wait' – a demand that the four Dreadnoughts to be ordered in that year should be increased to eight. Given the new German threat, the two-power standard was replaced by a requirement for a 40 per cent superiority over Germany in Dreadnought numbers.

The pace of construction did indeed increase – six Dreadnoughts were laid down in Britain in 1909 and six more in 1910, comprising the Neptune, Colossus, Indefatigable, Orion and Lion classes. Three Dreadnoughts being built in Britain for Turkey, Argentina and Chile were requisitioned by the Royal Navy on the outbreak of war in August 1914,[20] and by 1920, thirty-five Dreadnought battleships and thirteen battlecruisers had been built at a cost of £151 million by two royal dockyards (including nine at Portsmouth, see table above) and ten private yards.

But until peace, the storm

The darkness and the thunder and the rain.

Charles Sorley (1895–1915)

In February 1914, the additional labour requirements for the dockyards in the event of war had been estimated at 8,000–11,000, plus about 5,000 labourers for coaling and 1,200 men, mostly labourers, for the Naval Ordnance Department. In late July, the Board of Trade (BoT) was tasked with finding the additional skilled men (it was considered that labourers could be recruited locally without problem), and correspondence between the BoT, Admiralty and Treasury ensued.[1] At this time, shipwrights, boilermakers, coppersmiths, fitters, founders and patternmakers received 38s per week in the dockyards, other skilled trades between 28–37s, skilled labourers 26s and labourers 23s.

Recruiting Extra Labour

To facilitate recruitment, the BoT on 29 July suggested enhanced rates, e.g. an additional 10s per week to men recruited from a distance (or the London rate if higher), plus bonuses of 20 per cent for the first three months and 10 per cent thereafter (payable on completion), free board and lodgings, as well as railway fares and subsistence to those attending for interviews. The Admiralty would not agree to enhanced rates, but for those employed from a distance a subsistence allowance of 20s per week would be paid for food and lodgings if government accommodation and rations were not provided. Men who stayed for at least three months would get a 10 per cent bonus on wages paid, and a further 5 per cent for any subsequent period, on completion, unless they agreed to transfer to dockyard terms and establishment (assuming work was available).

The special emergency terms were agreed by the Admiralty on 1 August and the BoT commenced recruitment, even though it was a bank holiday which made things difficult. On 4 August, they reported that they had sent twenty boilermakers down from the Tyne, large numbers of fitters from Brighton and the London district and a limited number of electrical fitters from Glasgow, saying, 'There should be no doubt about the complete suitability of them all, except possibly as regards those from the London district.' They also commented that any number of boilermakers could be supplied from the Clyde and Tyne, and there was an especially good supply of highly qualified marine engine fitters who were likely to be idle on the Tyne. On 2 August, the Admiralty wrote to the Treasury setting out the emergency terms, asking for 'their covering sanction to the action that has been taken'. However, on 11 August, dockyard superintendents were instructed that the emergency terms should be discontinued as soon as possible, at their determination. By 22 August, 449 men had been engaged on the new terms at Portsmouth, including seventy-four 'skilled labourers' (which was questioned by the Treasury, and over whom there was some confusion).

The Treasury was not happy with the scheme, and wrote back on 24 August asking dockyards to make every endeavour to discontinue the emergency entry system. It felt that the special rates would engender dissatisfaction amongst regular staff, and in certain instances this had already shown itself. The dissatisfaction was offset only by the large availability of overtime. Men engaged under the special terms should be offered dockyard conditions with a guarantee of at least three months' work (subject to continued efficiency and good behaviour). If they refused, they could not be prevented from carrying on for the full three months on the emergency terms, but wherever possible new emergency entries should be stopped. A day later they relented, saying that they gave sanction 'for the present' for shipwrights and coppersmiths, especially the latter, to be recruited on the special rates, but the arrangement should 'cease without any delay as soon as circumstances admit'. Coppersmiths, though, might be obtained from Harland & Wolff, Belfast, where

PLACING H.M.S "MONARCH" IN FLOATING DOCK.
PORTSMOUTH. 3ᴿᴰ STAGE IN DOCKING. SILK.3.

The battleship *Monarch* in the floating dock. The dock had arrived in Portsmouth in 1912 and was then the largest floating dock in the world. It was moved to Invergordon at the start of the war, returning in 1919. (Portsmouth Royal Dockyard Historical Trust)

learn.[3] Volunteers from Australia and Canada were a further source of skilled labour at Portsmouth.[4]

Wages in the Yard

Weekly wage rates in January 1915 for dockyards in Great Britain

Classes of Workman	Hired	Established
Shipwright; Fitter; Boilermaker; Coppersmith; Founder; Patternmaker	38s	36s
Joiner; Plumber	36s	34s 6d
Smith	36s minimum	34s 6d minimum
Sailmaker	32s 6d	31s
Ropemaker	29s minimum	30s minimum
Skilled Labourer	24s	24s
Labourer	23s	24s
Rigger	30s 6d	29s

they were being laid off. Shipwrights might have to be guaranteed six months' work to obtain them on normal dockyard terms.

A number of cases were reported of emergency men being inferior to ordinary dockyard men, e.g. through inefficiency, intoxication or being AWOL, leading to them being discharged (fourteen at Sheerness alone). On 10 September, some men were still being recruited on special terms and dockyard superintendents were asked to report on the prospect of bringing it to a close.

Immediately after the outbreak of war, levels of overtime for dockyard employees increased sharply, with overtime pay rising from an average of about 3s 6d per man week to about 15s per man week in the main home yards. This was significant given that, at this time, the average wage for ordinary time was 32s 8d per week.[2] Some of the younger members of the workforce decided to enlist in the armed forces, which exacerbated the labour shortages.

Large numbers of women were employed in industrial roles in the dockyards during the war, often in semi-skilled tasks where modularisation and mechanisation of work (frequently using mechanically operated hand tools) compensated for the women's lack of skill in trades that often took a male apprentice six years to

In January 1915, wage rates were as shown in the table.[5] Established workers had guaranteed employment and qualified for a pension at the age of 60, whereas hired men has less job security and no pension.[6] The percentage of established workers was increased from 17.2 per cent in 1913 to 22.9 per cent in 1914, as the Admiralty, faced with increasing trade union consciousness, sought to emphasise the distinctiveness of dockyard employment when compared with private yards, which did not offer anything comparable to the established status.[7]

Employee petitions for increases were considered in January 1915 by the Admiralty, which investigated rates of pay in the commercial yards and made proposals to the Treasury for increases in pay of between 1s and 2s per week to all manual grades. However, the Treasury did not accede to these. In March 1915, the Admiralty referred the workers' claims to arbitration by the Committee on Production and Wages of Government Employees, and increases were awarded. Grades above ordinary (unskilled) labourers received an extra 3s per week, ordinary labourers and adult female employees 2s per week and apprentices, boys and girls 1s per week: in most cases this was more than the Admiralty would have conceded had the Treasury not intervened. Improved premiums for overtime and night shift pay were also awarded.[8] In addition to these pay awards, there were further significant wage increases in June 1916, August 1917 and April 1919 following arbitration. A war bonus was introduced on 1 April 1917, of 5s per week for male adults and 2s 6d per week for boys.[9]

In April 1919, skilled labourers had their minimum wage increased from 25s to 27s per week, with a maximum of 37s per week: this large range reflected the very varied work of skilled labourers. The maximum wage of 37s for hired men (or 35s for established men) could be achieved after five years if the employee was judged to be 'fully competent', and had been awarded in April 1917 (previously being 31s). Titles were also given to skilled labourers' occupations, viz: riveter, caulker, driller, welder, wireman, sawmill man, foundry furnace man, slinger, stage maker and ship-plate machine man. Claims for an increase to unskilled labourers' wages were not recommended and the minimum remained at 24s per week: however, a concession was that 15 per cent could be paid 25s, and 10 per cent could be paid 26s. By this time, it appears that all adult manual dockyard employees were in receipt of a war advance of 28s 6d per week and a bonus of 12.5 per cent, since these elements were recorded for patternmakers and all labourers at least.[10]

In January 1919, the wage of women clerks in the dockyards was increased by the Admiralty from 20s to 25s per week, in addition to which they also received 9s per week war bonus. This earned the Admiralty a rebuke from the Treasury, who had not been consulted and were concerned that women clerks in the War Office and Ministry of Munitions would lodge claims for a similar increase.[11] In 1920, a Whitley Committee was introduced at Portsmouth Dockyard:[12] this was a statutory council at which employer and trades union representatives met in a consultative forum which helped determine wage rates and terms and conditions of employment, and could take matters through to arbitration if necessary.

In the 1918–19 Navy Estimates,[13] the projected cost of all dockyard personnel was £13.84 million (of which £1.43 million was for overseas yards), more than treble the amount budgeted for in 1914–15, £4.02 million.[14] However, inflation was rampant during the war, totalling 97 per cent between 1914 and 1918:[15] in this light, the continual claims by dockyard employees for wage increases during the war can be more readily understood. The increase in dockyard personnel expenditure, adjusted for inflation, was 70 per cent. The average pay per employee (in both home and overseas yards, including salaried staff and police) increased from £73.87 in 1914–15 to an estimated £148.26 in 1918–19.[16] When adjusted for inflation, this represents an increase of only 1.9 per cent in real terms. Thus dockyard employees' earnings appear to have kept pace with inflation, and in fact fared better than the UK average earnings, which saw a reduction of 9.1 per cent in real terms between 1914 and 1918.[17]

The total workforce at Portsmouth dockyard had grown from 5,892 in 1880 to 11,924 by 1904, but reductions totalling 1,430 were made in 1905 and 1906. Thereafter, the workforce grew again to 14,736 by 1913 and to 16,287 in July 1914.[18] The latter figure included 15,646 male and fifty female manual workers and 591 officers and clerks. The workforce grew further during the war to 25,398 employees in December 1918, including 22,509 male and 1,786 female manual workers, and 1,103 officers and clerks.[19] The total wages bill rose from £1.16 million in 1913–14 to £3.67 million in 1918–19; in the same period, overtime payments rose from £69,000 to £807,000. The largest weekly wage bill was in the week ended 12 January 1918.[20]

Workload in the Yard

One aspect of dockyard work that had grown sharply during the Edwardian period was electrical engineering. In 1895, the first electrical workshop was opened at Portsmouth, and by the next decade electrical equipment was rapidly being adopted in both ships and the dockyard itself. *Dreadnought* had electrical fans for the engine and boiler rooms, as well as electric pumps, hoists, capstans, telephones and many other machines.[21] In the dockyards, a scheme executed between 1903 and 1908 electrified all lighting and machinery (except steam hammers and similar appliances).[22] The Electrical Engineering department at Portsmouth grew from 435 personnel in 1903 to 1,965 in 1919.[23]

The planned workload of the yard during 1913–14 was detailed in the Navy Estimates for that year.[24] New construction work included the battleships *Iron Duke* – which was to complete fitting out – *Queen Elizabeth* – to be launched – and *Royal Sovereign*, whose keel would be laid, and a depot ship was to be converted for destroyers. Under the heading of 'large repairs etc' were the battlecruiser *Invincible* (which was to receive numerous modifications) and the cruisers *Donegal*, *Doris*, *Isis*, *Kent*, *King Alfred*, *Pathfinder*, *Sentinel* and *Skirmisher*, all of which were to have their boilers retubed as well as other work. The royal yacht *Victoria and Albert*, listed for boiler retubing, and several smaller vessels were also included in the programme. Large repairs had a planned expenditure of £177,650, showing an increase of 78 per cent since 1905–06. The other main category, 'ordinary repairs', were budgeted at £486,250, an increase of 52 per cent over the 1905–06 figure. Included in the 1913–14 ordinary repairs plan was work on sixty-three named vessels, plus submarines, gunboats, small tenders, tugs and yard craft.

The actual work undertaken did not always conform with the plan set out in the programme of work, and this was the case at Portsmouth in 1913–14: the refits of *Donegal*, *Doris* and *King Alfred* were postponed, whilst large repairs were carried out on the battleship *Zealandia* and the cruiser *Blenheim*, which had not been provided for in the programme of work.[25]

After the outbreak of war, the demands on the yard increased dramatically, preparing ships for active service, supporting the

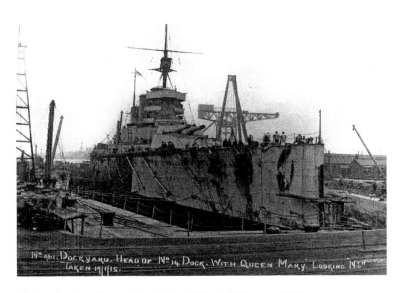

The battlecruiser *Queen Mary* in No. 14 Dock, 19 January 1915.

mobilisation of the reserve fleet, refitting old ships and equipping the merchant ships requisitioned for service with the fleet. This resulted in a big increase in the hours worked through overtime, including Sundays, and there was a need for more staff throughout the yard. A general order of 3 September 1914 at Portsmouth sought to share some of the labouring work in the yard with ships' companies. Work in connection with docking HM ships was to be performed by ships' companies as far as possible. The assistance to be rendered by naval ratings included making stages, scrubbing and cleaning ships' bottoms, scraping bottoms when the whole of the composition was to be removed, cleaning the dock and clearing it of debris, assisting in hauling ships in and out of dock and 'such unskilled labour as may be efficiently performed by them'.[26]

Seventy Belgian refugees were employed in the yard, helping to replace men who had enlisted in the Hampshire Regiment. A total of 1,385 men were lent to other yards (including the new dockyard at Rosyth), and 2,780 workers enlisted in the armed forces.[27] These losses were partly replaced by women, and partly by skilled men recruited from around the country and from Canada and Australia.

Women had been employed in the dockyard before the war in small numbers, in flag (colour) making and as clerks.[28] In August 1914, a total of fifty women were employed in manual work at Portsmouth.[29] In 1915, more women were recruited into clerical positions,[30] but the need for more industrial employees led to the infiltration of women into the traditional male domains of the Constructive, Engineering and Electrical departments, despite protests by the trades unions; for example, by 1916 there were 105 women in the Engineering department.[31] The women did not receive equal pay with men, and the Admiralty did not even abide by a minimum wage of £1 for women substituted in men's work which was recommended by the Labour Supply Committee.[32] Widows of chief petty officers and leading seamen and 'respectable orphan girls' were initially eligible for employment, but a few months later this was extended to all war widows and finally all women.

The women often received a hostile reception at first, suffering abusive taunts and jostling, but soon adapted to the rather unwelcoming environment and began to take on a wide range of tasks.[33] In the boiler shop, they were engaged in acetylene welding, in cleaning, pickling, galvanising and testing boiler tubes, and on lathes and drilling, screwing, punching and shearing machines, and

forges. In the Factory, they were employed in the working of lathes, planing, shaping, milling, engraving, buffing and slatting machinery; cleaning, cutting and testing condenser ferrules; cleaning air bottles for submarines and ships; and assisting the mechanics in cutting blades for condenser turbines. In the Gun Mounting Shop, they were employed on drilling, shaping, screwing and engraving machines, and also in fitting and adjusting range finders. In the Coppersmiths' Shop, they did acetylene welding, repairing lamps for ships. In the Pattern Shop, they were employed in making packing cases, painting patterns, turning wood plugs for boiler tubes and rollers for drawings, and also on the band saw machines. In the Drawing Office, they did the tracing of drawings and worked printing apparatus. In addition, women were employed in driving capstans for the movement of ships in basins, driving motor lorries, making overalls and flags, electrical wiring on board ships, sharpening saws and assisting at the machinery in the sawmills.[34]
Unlike the men, women industrial workers wore standard uniforms, with an ankle-length brown skirt and a brown tunic, and caps and triangular badges bearing the words 'On War Service'; but the skirts were not always practical and some switched to wearing trousers instead. Most took to wearing boots to protect their feet from the cobbled surfaces of the dockyard. Beatrice Hobby, who worked in the Block Mills – operating a band saw and painting pulley blocks – and on repairing Carley floats, had lost her husband, who was gassed in Flanders early in the war, leaving her with a six-week-old baby, so she had to work to support the child. Of her badge, she said, 'I was proud of it, to wear a badge for doing war work … I used to think it was wonderful that women could go and do such hard work, because it really was hard work.' She reported for work by 7 a.m., and sometimes would work on overtime until 8 p.m. It was a six-day week, with Sunday off.[35]

By 1917, the Constructive Department at Portsmouth employed 406 women, and the Engineering Department 247 women, out of a total of 1,750 women in the yard as a whole.[36] In 1918, the Engineering Department employed 346 women, out of its total workforce of 4,651. At its peak, 156 women were employed in the Sail Loft, as well as 110 men, making canvas gear, flags and protective clothing, and repairing life belts. The Electrical Department employed over 400 women. Dockyard office staffs were gradually increased by the entry of temporary male clerks and

Women workers in the Constructive Department, where, in 1917, 406 were employed. (Portsmouth Royal Dockyard Historical Trust)

Women carried out semi-skilled work on a wide variety of machines, as here working on parts for a K Class submarine on a horizontal drilling machine in No. 1 Shipbuilding Shop (which had been completed in 1867 as the Armour Plate Shop). (Portsmouth Royal Dockyard Historical Trust)

later by women clerks. In October 1918, the number of women employed in the yard was given as 2,122, comprising 1,842 manual workers and 280 female clerks.[37]

The workload of the yard during the war was summarised in a report produced in May 1919.[38] The battleships *Queen Elizabeth* and *Royal Sovereign* were under construction in August 1914, and were completed in January 1915 and May 1916 respectively. Besides carrying out the dockyard part of this work, much assistance was given by the yard to the machinery manufacturers to speed the completion of the two ships. Whilst the launch of the *Queen Elizabeth* on 16 October 1913 had been watched by 60,000–70,000 spectators, and was accompanied by all the ritual and panoply of a battleship launch in peacetime,[39] that of the *Royal Sovereign* on 29 April 1915 was a less ostentatious affair. Shipwright Edward Lane described the launch:

The submarine K2 was launched at Portsmouth in October 1916. She was driven by steam turbines on the surface and electric motors when submerged. Her class was not a success, with several lost in accidents, but K2 survived the war to be scrapped in 1926.

> On the day of the launching crowds assembled to see the sight. The chief constructor escorted the dignitary who was to perform the ceremony to the platform at the bows, the champagne bottle was broken and another ship was launched. Slowly at first but gradually gaining momentum she slipped into the waters of the harbour until the drag chains brought the ship to a standstill. A wonderful occasion for the spectators, but for the dockyard officers an anxious time. The day before the launching gangs of shipwrights from all over the dockyard assembled with 7lb mauls. They lined up along the length of the ship stem to stern to drive in the slivers (large wedges) in order to take the weight off the building blocks on which the ship was built and on to the sliding ways. Before the ship could move some of the blocks were removed by knocking out the steel wedges built into each set of blocks. The ship at this stage was in a very delicate situation – in other words, raring to go. When all was ready and the ship going down the ways, the watermen of Portsmouth, Stamshaw and Gosport would be paddling about like vultures ready to descend on their prey. For with the launch small baulks of timber or slivers of beech were by no means unconsidered trifles.[40]

After this, the last battleship to be built at Portsmouth, followed the construction of the submarines *J1, J2, K1, K2* and *K5*, which were laid down between April and November 1915; the first, *J1*, was completed in exactly twelve months, and the last, *K5*, in May 1917. The machinery of both of the J-class boats was assembled and installed in the yard. An order for a sixth battleship of the Queen Elizabeth class, to be named *Agincourt*, had been placed with the dockyard in 1914, but this was cancelled after the outbreak of war, as the emphasis on capital ship construction shifted to battlecruisers. The cruiser *Effingham* was laid down in April 1915, but the work was not fully manned owing to more important demands on other ships, and she was still on the slip when the war ended. Work was also carried out on four new ships received from contractors, *Courageous, Renown, Repulse* and *Emperor of India*: this in connection with incomplete items of work, alterations and additions.[41]

The battlecruiser *Princess Royal* was greeted by a noisy and jubilant welcome from every siren in the port when she arrived for repairs to damage sustained in the Battle of Jutland. She had been hit by one 11in eight and 12in shells from *Derfflinger, Markgraf* and *Posen*, twenty-two of her crew being killed and eighty-one wounded. Numerous fires were started and two legs of the tripod foremast were badly hit, but the ship remained operational.[42] Many other large repairs were carried out during the war, the most important of which were the destroyers *Nymphe* and *Ferret* (reconstruction of after part after depth charge and mine damage respectively), *Ettrick* and USS *Shaw* (reconstruction of fore parts); large torpedo damage repairs to the merchant ships

Clan Mackenzie (which also required new engines), *Limeleaf* (RFA), *Comrie Castle*, *PetinGaudet*, *Huntscape*, *Raranga* and others; and large repairs to the damaged bulges and main structure of the monitors *Erebus* and *Terror*. Other large refits and extensive machinery repairs were carried out to *Invincible*, *Kent*, *King Alfred*, *Suffolk*, *Carnarvon*, *Glory*, *Glasgow*, *Commonwealth*, *Zealandia*, *Vindictive* and others, and large collision damage repairs to various destroyers. The boilers of twenty-five big ships, forty-six destroyers and eleven torpedo boats were re-tubed or partially re-tubed, and four boilers were manufactured for the tug *Roysterer*, which was built at Southampton by Thornycroft.[43]

Conversions of ships for special services included the former mercantile *Aro*, *Sobo*, *Sokoto* and *Diligence* as depot ships; *Racer* and *Mariner* as salvage vessels; and *Courageous* and *Amphritite* as minelayers. Bulges were fitted to the battleships *Commonwealth*, *Revenge* and *Resolution*, and there was constant rearming of monitors. Various merchant ships were armed and a number of Q ships were fitted out; yachts, trawlers and drifters were fitted out for the auxiliary patrol; and 576 motor launches were received from America, armed and trials carried out.[44] (These launches were 80ft long, except the first fifty which were 75ft, and were built by Elco of Bayonne, New Jersey. They were shipped to Britain as deck cargo.[45])

After the launch of a battleship at Portsmouth, local watermen collect the tallow which had eased the ship's progress down the ways. (Portsmouth Royal Dockyard Historical Trust)

Royal Sovereign, completed in May 1916, was the last Portsmouth-built battleship.

The Constructive Department reported that 1,140 vessels underwent ordinary refit, for a period of one week or more, during the war, including, *inter alia*, thirty-four battleships, four battlecruisers, thirteen cruisers, nine light cruisers, 408 destroyers, 148 torpedo boats, thirty-one monitors, 137 trawlers and drifters, twenty-three minesweepers, seven armed merchant cruisers and hospital ships, twenty submarines, twenty fleet auxiliaries, twenty-seven tugs and six Q ships. In addition, work on 1,056 vessels had been contracted out by the dockyard to local yards under the supervision of the dockyard officers. In March 1919, the manager of the Constructive Department of the dockyard reported that 1,658 warships had been dry-docked or hauled up onto slipways for repairs.[46]

The yard conducted a range of experimental work associated with the construction and trialling of inventions for the defence of ships against mines, torpedoes and submarines, including minesweeping gear such as paravanes, anti-submarine mines, mine plugs, depth charges, searchlights and hydrophones.[47] Attention to the port's defences involved work by the yard in the manufacture and placing of anti-submarine nets across the harbour mouth from Fort Blockhouse to Point, and mooring the hulks *Ant* and *Insolent* as gate vessels; for Spithead, an anti-submarine net was manufactured and placed between Horse Sand Fort and No Man's Land Fort, with the gate vessels *Argo* and *Asov*.[48]

The manufacture of guns, gun mechanisms and shells introduced a new element to dockyard work, starting in July 1915, and included 170 quick-firing 12 cwt guns and fifty-five 4in guns, which were manufactured from rough machined forgings, and about 10,000 shells were produced from October 1915 onwards.[49]

The naval stores department saw a dramatic increase in the amount of stores issued to the fleet and dockyard, the value of which increased from £5.8 million in 1913–14 to £23.1 million in 1917–18. The amount of oil fuel issued rose from 30,620 tons to 235,350 tons, whilst petrol increased from 236,350 tons to 801,214 tons, and issues of coal and coke doubled.[50]

There was a great deal of military supplies (guns, ammunition, tanks, locomotives, vehicles and stores) and 21,166 troops (officers and men) embarked in the dockyard in merchant ships for transport to France. This involved 15,292 loading lifts and 3,184 unloading lifts by the 250-ton crane, and many by the 25-ton electric crane. Three million shells were loaded by electric cranes. Although all of the military materials and troops were delivered safely to France, two of the transports were sunk by mines, in both cases the ships being light.[51]

The larger dockyard tugs, *Swarthy*, *Sprite*, *Grappler*, *Pert* and *Empress*, were on twenty-four-hour standby to be sent from the harbour to deal with the salvage of vessels in distress or damaged, and were called upon on 304 occasions. Seventy-seven of these vessels were safely brought into port (some being first beached at Stokes Bay) and a large proportion of the remainder, after running aground, were refloated with the aid of dockyard tugs. The salvage of a seaplane which had fouled one of the wireless masts at Horsea Island in September 1917 was accomplished by the yard's riggers.[52]

When the war ended, many of the women workers were laid off, so by April 1919 their numbers had reduced to 460, and many of the men who had enlisted returned to the yard.[53]

Other Incidents

On 25 September 1916, Commander Heinrich Mathy and his crew climbed into the gondola of the German Zeppelin *L31* and weighed off, heading for England. His main target was London, but upon reaching Dungeness the night sky was clear and he decided that an attack on the heavily defended capital would be difficult, and instead he would bomb Portsmouth dockyard, where, as Mathy pointed out to his crew, 'nobody has ever visited and it is sure to be very interesting'. Mathy attacked from the south, but was picked up in the searchlights of HMS *Vernon* at 11.50 p.m., and the batteries of the port and warships in the harbour let off a barrage of anti-aircraft fire. One of Mathy's crew members recorded: 'Dozens of searchlight clusters find us and fix on us. An unearthly concert is unleashed conducted by Satan himself.' This probably referred to anti-aircraft fire from Point battery and Whale Island as *L31* came over the harbour. A heavy bomb load of 8,125lb was dropped, and Mathy, blinded by the searchlight, reported that 'all bombs has fallen on the city and the Dockyard'. However, the bombs appear to have fallen harmlessly into the harbour and there is no record of a hit on the dockyard. Six days later, *L31* was attacked by a fighter pilot, Second-Lieutenant W.J. Tempest, over Potters Bar in Hertfordshire: Mathy chose to leap to his death from the plunging inferno, whilst his crew perished in their craft.[54]

On 11 January 1918, a skilled labourer in the dockyard, John Lomax, aged 48, was sentenced to six months' imprisonment with hard labour on a charge of making statements likely to cause disaffection to the king. It was alleged during the dinner hour in the dockyard dining hall that he had made an excited pro-German statement, in the presence of fellow workmen, among whom was a wounded soldier. They resented his conduct and gave information to the authorities.[55]

When the ship that is tired returneth,

With the signs of the sea showing plain,

Men place her in dock for a season,

And her speed reneweth again.

Rear admiral Ronald A. Hopwood RN (1868–1949),
The Laws of the Navy

Retrenchment

The mood of the major nations, horrified by the slaughter of the Great War, drove a post-war imperative to avoid another arms race. At the Washington Naval Conference of 1921–22, a ten-year agreement fixed the ratio of battleships at 5:5:3, i.e. 525,000 tons each for the USA and Britain, and 315,000 tons for Japan, whilst smaller limits with a ratio of 1.7 applied to France and Italy. Battleships could be no larger than 35,000 tons, and the Royal Navy had to scrap or demilitarise twenty-four of its Dreadnought battleships and battlecruisers to comply with the overall tonnage limits of the treaty. For aircraft carriers, Britain and the USA allowed themselves 135,000 tons each, whilst Japan had 81,000 tons, and some were converted from battleships. Even with the Washington Treaty, the major navies remained suspicious of each other, and engaged in a race to build cruisers which had been limited in size (10,000 tons) but not numbers.[1] That loophole was closed by the London Naval Treaty of 1930, which specified tonnage limits for cruisers and destroyers. The number of heavy cruisers was limited – Britain was permitted fifteen, with a total tonnage of 147,000, the US eighteen totalling 180,000 tons and the Japanese twelve totalling 108,000 tons. For light cruisers, no numbers were specified but there were tonnage limits of 143,500 tons for the USA, 192,200 tons for Britain and 100,450 tons for Japan. For the first time, submarines were also limited, Japan being given parity with the US and Britain at a total of 53,000 tons each.[2]

There was an inevitable contraction in the workload at Portsmouth Dockyard and in the numbers employed, due to the reduced requirements of the post-war fleet. The number of industrial grade employees fell from 24,295 in December 1918 to 16,269 in March 1920, a reduction of 33 per cent.[3] The position was exacerbated by the need to re-employ dockyard workers who had joined the forces in the war and been promised a return to work when the war ended. Many workers had to be laid off, including women, pensioners, men called in from other areas and occupations and volunteers recruited from the empire. In 1921, the Admiralty ordered a reduction in the working week of seven hours so that approximately 1,400 of the men discharged could be re-employed, but this scheme only lasted for a year before the men were again discharged. By March 1923, numbers in the yard had fallen to 11,473 as a result of the 'Geddes Axe', an initiative to cut the national debt through reductions in government expenditure, steered by the Committee on National Expenditure which was chaired by Sir Eric Geddes, cutting defence expenditure from £189.5 million in 1921–22 to £111 million in 1922–23.[4] The country was blighted by a depression, with high levels of unemployment throughout the 1920s, and both the arms limitation treaties and a climate of austerity led to continual cutbacks in the size of the fleet.

In 1919, the large floating dock, by then known as AFD 5, which had been moved to Invergordon at the start of the war, was safely returned to Portsmouth and berthed in its old position at the north-east end of the yard, where it stayed until 1939. But the floating crane, known as No. 1 Crane Lighter, which had also been sent north, was wrecked on the east coast when returning under tow. A smaller floating dock was positioned in Haslar Creek for submarines.[5]

A large floating crane with a 200-ton lift was handed over by Germany to the Admiralty as part of war reparations. It was a duplicate of a crane used at Wilhelmshaven that had been erected in 1916. The pontoon and its propelling steam machinery had been built at Kiel by Deutsche Werke, whilst the electrically driven crane had been made by Deutsche Maschinefabrik A.G. (Demag). The pontoon was towed to Portsmouth in May 1923, and the crane parts

The huge floating crane was received from Germany in war reparations. It had a lift of 200 tons and remained in service until the late sixties.

1967, the crane was downgraded to lift only 60 tons when cracks were discovered in the structure (it was finally sold in April 1970 to the Portsmouth shipbreaker H.G. Pounds, and was apparently sold on, possibly for further use, leaving the harbour in June 1970 under tow for Greece).[8] A smaller floating crane with a 5-ton lift was also received under reparations.[9]

The cruiser *Diomede*, under construction by Vickers at Barrow, and in the process of fitting out, was towed to Portsmouth, where she was completed in February 1922. Similarly, the incomplete aircraft carrier *Eagle*, being built by Armstrong Whitworth on the Tyne, was steamed to Portsmouth under her own power on 24 April 1920 for completion. She finally completed sea trials in September 1923. These contracts brought work to the dockyard whilst freeing commercial yards for merchant ship construction. Work on the cruiser *Effingham*, laid down at Portsmouth in 1917, was revived, and she was launched on 8 June 1921 and finally completed in July 1925.[10] Unusually, to provide work for the yard, an oil tanker was built for Anglo Saxon Petroleum Co., London (which provided oil transportation for the Royal Dutch Shell group): she was the *Murex* of 8,887 tons, launched in 1922.[11] A sister ship, *Nassa*, was built at Devonport Dockyard in the same year.

In 1920, a Whitley Committee was introduced at Portsmouth Dockyard:[12] this was a statutory council at which employer and trades union representatives met in a consultative forum which helped determine wage rates and terms and conditions of employment, and could take matters through to arbitration if necessary. It helped formalise the negotiating rights which the Admiralty had finally conceded to trades unions.

In November 1923, there were 12,000 unemployed in Portsmouth and Gosport and a deputation led by the Portsmouth Local Employment Committee met with the Parliamentary Secretary of the Admiralty, seeking some amelioration of the situation and putting forward suggestions which might improve matters, including letting out part of the dockyard for private work, rebuilding the infrastructure of the dockyard, improving and covering the main slipway and ordering a new cruiser from the yard. They also sought assurances on the strategic future of the yard, in the light of its vulnerability to aircraft from the Continent in the time of war. In response, an undertaking to take on between 1,200 and 1,400 men in the dockyard and ordnance depots in the forthcoming winter was

were transferred by ship. The crane was to be erected by Messrs Cowans Sheldon with the assistance of German overseers, but the latter withdrew following a breakdown in reparations arrangements, so the whole of the erection and testing was undertaken by the contractors.[6] The whole structure weighed 4,000 tons. All motions of the crane itself were electrically operated, with power provided by two steam-driven turbo-generators of 270hp each situated in the pontoon hull. The jib alone weighed 190 tons, and the heaviest lift performed was 267 tons during a test lift.[7] This crane proved to be a valuable addition to the yard and remained in service until the 1960s. In 1942, it suffered a near miss from a German bomb that dislodged some of the counterbalance weights, with one of 15 tons being thrown into the air and falling back through into the engine room, whilst the detonation of the bomb in the water put great dents in the bottom plating. Fortunately there were no casualties. In

HMS *London* was one of three heavy cruisers ordered from the yard in the twenties, giving a much-needed boost to employment.

The Victory Gallery was completed in 1938 to house W.L. Wyllie's 42ft-long painting *The Panorama of the Battle of Trafalgar*, and Admiral Nelson and HMS *Victory* exhibits.

given. Although it was known that several new cruisers were to be ordered, no undertaking to place one with Portsmouth could be given.[13] About 1,100 men were recruited into the yard, and there was relief on 15 May 1924 when an order was placed to build the cruiser *Suffolk* at Portsmouth, giving work on the main slipway, which had been empty since June 1921. Orders were to follow for the cruisers *London* and *Dorsetshire*, which were laid down in 1926 and 1927 respectively. These measures gave an element of stability to the employment situation for the next three years. However, following lay-offs in 1927–28, the number of industrial employees fell further to 10,646, including 107 women, in March 1929.[14]

In 1926, the government announced the closure of Pembroke Dockyard and the reduction of Rosyth Dockyard to care and maintenance. Room at Portsmouth had to be found for several hundred established men from these yards, displacing hired men and causing resentment in the dockyard.[15]

Refitting and modernising the remaining battleships had become an important part of the yard's work. *Royal Oak* was in hand from September 1922 until April 1924, *Warspite* from 1924 until April 1926, *Resolution* from 1926 to 1927, *Queen Elizabeth* from May 1926 until January 1928, *Royal Sovereign* from 1927 to 1928, *Malaya* from September 1927 to February 1929 and *Valiant* from March 1929 until December 1930. The battlecruisers *Renown* and *Repulse* had major work to increase their armour and protection

from May 1923 until August 1926 and January 1919 to December 1920 respectively,[16] and *Hood* was refitted between May 1929 and March 1931.[17]

Rebuilding and modernising the yard after the war was constrained by the available finance, but nevertheless good progress was made in the 1920s. The Rigging House, Rangefinder Test House and Semaphore Tower, which had been destroyed by a fire in 1913, were rebuilt incorporating the original eighteenth-century foundations and the old Lion Gate, and completed in 1930.[18] The Lion Gate dated from 1778 and was originally the main entrance through the rampart fortifications to the town of Portsea. It had been moved to form the main entrance to the Army's Anglesey Barracks, which became the Navy's Victory barracks.[19]

The Main Electricity Generating Station was modernised to meet the much-increased demand for electricity from electric furnaces for the manufacture of steel castings, large electric air compressors which serviced pneumatic hand tools for caulking, riveting, chipping and drilling, electric welding, electric lifting appliances and the electrification of the Main Pumping Station. This work was completed in December 1928. In 1929, the generating station consumed 31,152 tons of coal.[20] There was major expenditure on the reconstruction in reinforced concrete of the Boat House Jetty, Pitch House Jetty and the South Railway Jetty, and the North West Harbour Wall.[21] A new oxygen plant, coppersmiths' shop, steel foundry and central estimating office were built in the 1920s, and a self-propelled 25-ton floating crane was delivered in 1928 by Cowans Sheldon. The Victory Gallery was built beyond the north side of No. 11 Storehouse in 1929–38 to house HMS *Victory* artefacts and associated museum displays after the ship was docked in No. 2 Dock in 1922.[22] Following conservation work, *Victory* was opened to the public on 17 July 1928 by King George V. In 1936, the Central Metallurgical Laboratory was established in the dockyard, providing valuable analysis and testing of metal components and failures.

Navy Weeks

In August 1927, the first Navy Week was held at Portsmouth, in aid of naval charities. The dockyard opened its gates between 1.45 p.m. and 6 p.m. each day to the general public, who paid 1*s* per adult and 6*d* per child to see and go aboard the warships on display. Visitors could go where they liked in the dockyard, though there was a suggested itinerary under which they were first shown around the Dockyard Museum and then given a tour of HMS *Victory* by naval ratings in Georgian period dress. They could then go aboard the battleship *Iron Duke*, the cruisers *Centaur* and *Curacao*, a sloop, a mine-laying monitor, the submarines *L6* and *L4* and a coastal motor boat, and tea could be taken on the South Railway Jetty to the accompaniment of music played by a naval band. There were queues to get aboard the submarines, always a popular attraction at such events. On the first day, despite frequent showers, the number of visitors was over 8,000, well above the 2,000 expected.[23] Over the six days, a total of 48,765 people paid for admission.

The success of the event meant that it was repeated the following year, on a larger scale: the battleship *Benbow*, battlecruisers *Renown* and *Repulse* and the aircraft carrier *Furious* headed the line-up of ships open to the public, and admissions totalled 89,583 – nearly double that of the previous year. The opening of *Furious* to the public provided the first ever opportunity for them to go aboard an aircraft carrier, and many of the visitors took advantage of what was probably the most popular attraction. Also open to the public were the cruiser *Centaur*, the new destroyer *Amazon*, the destroyer leaders *Wallace* and *Campbell* and four L-class submarines, and on their tour of the dockyard visitors could see the cruisers *London* and

Navy Week with HMS *Victory* and, in No. 1 Basin, L-class submarines.

Dorsetshire under construction. The civic authorities extended the event by illuminating the piers and the gardens at Southsea beach and providing seaside entertainment each evening after the dockyard closed.[24] Coach parties arrived from many parts of London, the Midlands and the southern counties, whilst some visitors arrived by train at the conveniently sited harbour station. Thereafter, Navy Week became an annual event, up to and including 1939. As the Navy became more inventive in devising added attractions, the popularity of the event grew, and a new record attendance of 206,759 was achieved in 1936.[25] In that year there was a display by Nimrod and Osprey aircraft of No. 800 Squadron (HMS *Courageous*) conducting a massed attack on the battleship *Iron Duke*, which 'dive-bombed and machine-gunned' the ship at close range; a demonstration of a night action between two warships; a display by the new motor torpedo boats, showing how they could be manoeuvred as a complete squadron at high speed; a demonstration of the battles of Coronel and the Falklands using ship models moved by *Victory* cadets across a stage set to represent the South Atlantic and Pacific; and the big ship attractions included the battlecruisers *Hood* and *Renown* and the aircraft carrier *Courageous*.[26] As the rearmament programme gathered pace, there were many new warships on view: in 1939, new ships scheduled to attend at Portsmouth included the aircraft carrier *Ark Royal*, the cruiser *Belfast* and the destroyers *Somali*, *Janus*, *Javelin* and *Jersey*.[27]

Work in the Yard

In 1929 a concession was announced, giving dockyard employees six days' paid leave each year. From 1930 this leave was taken during the first week in August, and whilst the Dockyard was closed Navy Weeks were held.[28] On 12 September 1930 the keels of three vessels were laid in No 13 Dock. They were the destroyers *Crusader* and *Comet* and the mining tender *Nightingale*, all launched on 30 September 1931.[29] Two steam dockyard tugs were built in 1932 for service in Malta – they were *C307* and *C308*, each of 192 gross tons.[30] Construction of the mining tender *Skylark* (launched 1932) and the destroyers *Duncan* (1932) and *Exmouth* (1934) followed.[31]

The destroyer HMS *Crusader* was launched at Portsmouth in September 1931 and completed in May 1932, and in 1938 was transferred to the Royal Canadian Navy to become HMCS *Ottawa*, seen here.

Extensive reconstructions and modernisations of capital ships provided much needed work for the yard in the mid to late 1930s. In March 1934 *Warspite* was taken in hand for a major reconstruction and modernisation. She received new engines and boilers, and 1,100 tons of extra armour. Her armament was updated and an aircraft hangar, cross-deck catapult and two cranes were fitted. The work was completed in March 1937 at a cost of £2,363,000.[32] Her sister ship *Queen Elizabeth* was then reconstructed, between August 1937 and December 1940, when she ran initial trials and moved to Rosyth for completion of the work.[33] The battlecruiser *Repulse* was modernised between April 1933 and May 1936, with improved armour protection, updated secondary and anti-aircraft armament and the addition of a hangar, cranes and catapult for two aircraft.[34] Her sister ship *Renown* followed between September 1936 and August 1939, with new engines and boilers, added armour, updates to her armament and gunnery control equipment, and provision for aircraft.[35]

Meanwhile, new build work on cruisers resumed, with *Neptune* (launched 1933), *Amphion* (1934), and *Aurora* (1936). As a rearmament programme gathered momentum industrial employee numbers rose by 1,611 between April 1934 and March 1936. By March 1939 a further 2,680 had been added, bringing the industrial workforce up to 15,036.[36] The dockyard budget grew from £2.7 million in 1934–35 to £4.4 million in 1937–38. Portsmouth

The dockyard departments usually held Christmas parties for the children of the employees. Here the welders' and burners' Christmas party on 7 January 1939, held in the Drivers' Café on The Hard, is shown. (Portsmouth Royal Dockyard Historical Trust)

remained the biggest British dockyard, responsible for 30 per cent of all dockyard work.

After a two-year hiatus during which there had been no ship under construction, the keel of the cruiser *Sirius* was laid on No. 5 Slip on 6 April 1938.[37] In 1937 work had started on building a new No. 4 Boathouse in place of the old structure (which dated from about 1700). The northern end was demolished and the first half of the new building was erected. However, the war then started and the southern exposed and unfinished end was quickly given a temporary corrugated iron sheeting end wall. The building remains in that state to this day.[38]

Let him who desires peace, prepare for war.

Vegetius, fourth century, Epitoma Rei Militaris
bk. 3 Prologue.

Dockyard Activity

At the start of hostilities in September 1939, the threat of attacks on Portsmouth by German bombers was recognised and had implications for the work that could be undertaken in the yard. The risk to precious capital ships was soon considered to be unacceptable, and in most cases these had to be taken elsewhere for work. The battleship *Queen Elizabeth* was undergoing a major modernisation in the yard, having been taken in hand in August 1937. Amongst many other changes, she was fitted with new turbines and boilers and a new secondary armament of twenty 4½in guns replacing her 6in guns, with their casemates plated over. Aircraft hangar stowage was provided abaft the funnel for four Walrus aircraft and an athwart ship catapult fitted. The bridge structure was completely redesigned as a slab-sided tower citadel, and a tripod mainmast was rigged. She was in the process of final fitting out and trials, but because of air raids she left Portsmouth in December 1940 to complete her final trials at Rosyth in January 1941. Emerging from this comprehensive makeover, she was to play a frontline role in the fleet, as were the others of her class, in contrast to the newer but less extensively modernised Royal Sovereign class which played a more secondary role.

The battleship *Nelson* was in C Lock from 15 January to 25 April 1940 for repairs to major structural damage sustained when she had been mined entering Loch Ewe on 4 December 1939. The damage was repaired at Portsmouth, but her refit was incomplete when she sailed from the port on 6 June 1940 to avoid the threat of bombing, and the work was completed at Greenock.

The aircraft carrier *Ark Royal* was in D Lock from 19 February to 1 March 1940 for recoating her bottom; in the accompanying short

The Portsmouth-built battleship *Queen Elizabeth* during the war, following her second modernisation at Portsmouth Dockyard. Her early war service in the Mediterranean was curtailed by the heavy damage she sustained from an Italian manned torpedo attack in Alexandria harbour. After repairs in America she became flagship of the Eastern Fleet.

refit, she was fitted with degaussing equipment (which reduced the magnetic field of the ship, making it less vulnerable to magnetic mines). The battleship *Resolution* was sent to Portsmouth for repair and refit in March 1941 following damage sustained when she was torpedoed off Dakar. But little work was done because of the bombing threat, and she left Portsmouth for the USA instead.[1]

The workload in the dockyard had been heavy since 1936, and intensified further. The yard's dry-docking register shows the following numbers of vessels docked during the war:[2]

1939 243
1940 271
1941 195
1942 251
1943 242
1944 877
1945 286

The huge peak in 1944 was caused by Operation Neptune, the naval component of Operation Overlord, the Normandy landings. In total, 2,365 vessels were repaired or refitted in dry docks during the war years.

The floating dock AFD 5 left Portsmouth for Alexandria on 24 June 1939, arriving there on 20 July,[3] and was replaced by AFD 11, which had been built in 1924 for the Southern Railway's Southampton Docks and was the biggest in the world.[4] New construction continued in the yard during the war. The cruiser *Sirius* was launched on 18 September 1940 by Lady James, wife of the Commander-in-Chief, Portsmouth. During fitting out she suffered bomb damage from near misses, but was completed on 6 May 1942. Her place on the slip was later taken by the keel of the cruiser *Hawke*, laid down on 1 July 1943. Due to other priorities, she was incomplete at the end of the war and her hull was broken up on the slip.

Two floating docks were built on No. 5 Slip, AFD 18 being completed in 1942 and AFD 21 in 1943. The submarines *Tireless* and *Token* were laid down in No. 13 Dock on 30 October 1941 and 6 November 1941 respectively. They were both floated out on 19 March 1943 and completed in 1944. They were replaced in the dock by the submarines *Tiara* and *Thor*, which were laid down in

April 1943 and floated out a year later. Incomplete at the end of the war, they were both scrapped.[5] A midget submarine, *X4*, was assembled in No. 4 Boathouse in 1942–43: the bow was built at Hull, the stern at Devonport and the control room at Portsmouth.[6]

Boom defences were laid to protect the harbour from surface attack. One boom stretched from Lumps Point to Nettlestone Point, a second from Durns Point across the Needles Passage to Hamstead Ledge and the third was held in readiness at the Round Tower to be hauled across to Fort Blockhouse to block the harbour entrance in an emergency.[7]

As in the First World War, women were recruited into the dockyard in much increased numbers, and the work they undertook was usually that of skilled or unskilled labourers (both being industrial grades) or clerks. The women industrial employees were in jobs such as welding or light machining. At first recruitment was voluntary, but this did not produce enough recruits and in March 1941 it was made compulsory, as a form of conscription, part of a wider national mobilisation of women into industry and the forces.[8] Women with children under 14 did not have to work, although some did. In March 1939, there were 154 women employees in Portsmouth Dockyard, and a year later this had actually reduced by four. However, in the next twelve months 790 women were

The Portsmouth-built cruiser *Sirius* was completed in May 1942 and spent the war in the thick of action in the Mediterranean, apart from a stint supporting the Normandy landings.

recruited and the total number of women increased to 921 by March 1941. This peaked at 2,617 in March 1944. At the end of the war, women who wished to leave were allowed to do so, so that by March 1946 only 353 were employed. These figures appear to only include industrial grades.[9]

The total industrial workforce in the yard in March 1939 was 15,036. This peaked at 17,478 in March 1943, and by March 1946 had fallen to 15,645. The numbers of women leavers had been partly offset by returns of men from the forces.[10] In 1940, 339 apprentices were recruited at Portsmouth Dockyard, plus eighteen Royal Navy shipwright apprentices. About eighty-six of the 339 were indentured as shipwright apprentices.[11] One apprentice recruited in 1943 recalled that his indentures said he was not to 'contract marriage during the period of this indenture; nor be guilty by word or action of any immoral, indecent, irregular or improper conduct or behaviour in any respect whatsoever, but shall and will demean himself at all times with strict propriety and submission to his superiors'.[12]

In November 1943, the Main Gate (1711, later the Victory Gate), originally designed by the Master Shipwright, was widened from 12ft to 22ft to allow access to larger vehicles. The wrought-iron arch and lantern between the piers were removed.[13] Also in 1943, three parallel city streets – Marlborough Row, Gloucester Street and Frederick Street – were acquired and incorporated into the dockyard. Marlborough Gate, originally built in 1711, was moved to the apex at Bonfire Corner. The area taken in had been known as the Marlborough Salient as it protruded in to the Dockyard, and it had been planned to be taken into the Dockyard during the mid 1930s; clearing of the area by German bombing finally made this possible.[14]

Air Raids

The city of Portsmouth was a prime target for bombing raids by the German Luftwaffe, and officially suffered sixty-seven air raids between July 1940 and May 1944, three of these categorised as major attacks. The three major raids took place on 24 August 1940, 10 January 1941 and 10 March 1941, and in each case bombs fell onto the dockyard. In addition, there were nineteen other dates between 11 July 1940 and March 1941 when bombs fell onto the dockyard. Although locations such as the yard and other military installations were undoubtedly the primary targets, aerial targeting at that time was not particularly accurate, especially during night-time raids when visual methods could not be used and other methods such as dead reckoning or radio triangulation were used instead. As a result, large numbers of civilian residential areas were hit. During the four-year period of the Portsmouth Blitz, 930 people were killed, 1,216 hospitalised and a further 1,621 suffered less severe injuries.6,652 properties were totally destroyed, 6,549 were seriously damaged and almost 69,000 houses suffered some form of minor bomb damage.[15]

As well as the air raids, there were 1,581 alerts, which caused almost as much disruption for they often stopped work in the dockyard for many hours at a time, as workers sought refuge in shelters until the all-clear was given. High absenteeism and late mustering also became problems as workers struggled to get to work through streets blocked with rubble and transport services were disrupted. The first air raid on the dockyard came on 11 July 1940, but little damage was done to the yard itself, though the city suffered much more.[16]

On Monday, 12 August 1940, a daylight raid on Portsmouth and the dockyard was made by about fifty-five Ju 88 bombers, accompanied by Bf 109 fighters, at around midday, in what was the first major air attack on a populated area of England. Some of the aircraft approached from the south and came down to a level of 6,000–7,000ft, above the barrage balloons, to attack. Others approached from the north, and when inside the balloon barrage came over Portchester Creek and delivered dive-bombing attacks. Forty-one bombs fell on naval establishments or in the harbour, whilst 130 other bombs fell in the neighbourhood, including seventy-two on the Isle of Wight. Gunfire from two 40mm Bofors guns and one 12-pdr gun, with Army crews, on Whale Island downed two or possibly three of the aircraft. Seventeen men were killed in the naval establishments, including ten RAF personnel who were part of the barrage balloon complement and were in a shelter at HMS *St Vincent* at Gosport. Also killed in the shelter were two civilian groundsmen, and nearby two able seamen (part of an anti-parachutist platoon on duty in Forton Field) were killed. In the dockyard, two people were killed in No. 11 Storehouse and one in the Rigging House. Nineteen people were seriously wounded. Many buildings and docks in the dockyard were damaged, including substantial damage to the Chain Cable Test House, No. 11 Storehouse, No. 1 Dock, the Factory, No. 12 Dock Shed, West End Torpedo Depot, the South Wall near the new Boathouse and C Lock. The estimated cost of repairs to the buildings and docks was £44,500. Three boats were sunk in the South Camber, including the Commander-in-Chief's barge and a rear admiral's barge. A trawler and a drifter in No. 1 Basin were damaged. There was further damage to HMS *Victory*, Crane Lighter No. 1 and the RN Barracks, and on the Gosport side to HMS *St Vincent* and HMS *Hornet*. Two dockyard tugs were sent to assist in fighting a large fire at Portsmouth Harbour station, whilst two dockyard fire-engines attended a large fire in the civil area near the dockyard.[17]

On Saturday, 24 August 1940 at 2 p.m., German bombers dropped sixty-five high-explosive bombs on the city. There were also a few incendiary bombs, and many more of these were later used in the night-time raids. In the dockyard, there was substantial damage to five sites – No. 33 Store, No. 7 Dock Gear Store and Dining Rooms, C Lock, the North-west Wall and D Lock Caisson Chamber. There was further damage to buildings at HMS *Vernon*. Twenty-six people were killed when a shelter in the dockyard took a direct hit, and another six people were seriously injured. The cost of repairs required to buildings was estimated at £28,500.[18] The destroyers *Acheron* and *Bulldog* were damaged whilst in the dockyard undergoing repairs to earlier bomb damage they had sustained in the English Channel. *Acheron* was lying at the North-west Wall and was hit by a bomb aft, which exploded within the stern part of the ship and caused major fire damage to machinery, Y gun and structure. Two people were killed and three injured. *Bulldog* was lying alongside *Acheron* and was damaged by splinters, which also killed her commanding officer, Lieutenant Commander John Wisden, RN. *Acheron* was speedily repaired by the yard, so that by 17 December she was running speed trials at sea south of the Isle of Wight. She struck a mine in a previously swept area and sank with heavy loss of life: 167 of her ship's company and twenty-seven dockyard workers perished, and there were only nineteen survivors.[19]

During an air raid on 5 December 1940, thirty high-explosive bombs were dropped on the dockyard: the destroyer *Cameron* was hit, set on fire and turned over onto her side, whilst under refit in No. 8 Dock. There was severe damage and the vessel was declared a constructive total loss. In the same raid, damage to buildings was

suffered at Royal Clarence Yard and HMS *Vernon*. A shelter was hit, resulting in one fatality and four people seriously injured. In a further raid seventeen days later, three boats were sunk in the South Camber with four casualties.

On the night of 10–11 January 1941, large fires were started in the dockyard by incendiary bombs, in Nos 4, 5, and 6 Boathouses, No. 14 Store, the Dockyard School, Cashier's Office, No. 3 Rigging House and the woodpile and timber store. High-explosive bombs damaged Nos 7 and 9 Docks, the main pumping station, No. 14 Dock caisson and the anti-aircraft ship *Tynwald*, which was undergoing conversion from an Isle of Man ferry into that new role. *Tynwald* suffered considerable structural damage but was subsequently repaired.[20] The big paint store was ablaze, becoming a veritable furnace, and the paint pots were said to be exploding like artillery fire. Part of the Officers' Block and one Seamen's Block in the naval barracks were burnt out.[21] This raid also caused extensive damage in the city including to the power station, causing a loss of electrical power, and power was supplied to the city from the dockyard power station until the main supply could be reinstated.

On 31 January 1941, Prime Minister Winston Churchill toured Portsmouth dockyard. *The Times* reported, 'Spontaneous cheering

No. 4 Boathouse was completed in 1939 and now houses a boatbuilding training centre and a display of historic small boats.

greeted him wherever he went, and he gave the "thumbs-up" sign to dockyard workmen who assured him by repeated shouts that they were not down-hearted.' Afterwards he visited the bombed area of Portsmouth and said:

> I have thought about you and your friends in Southampton a good deal when we knew how heavily you were being attacked, and I am glad to find an afternoon to come and see you and to wish you good luck and offer you the thanks and congratulations of the Government for the manner in which you are standing up to these onslaughts of the enemy.[22]

On the night of 10–11 March 1941, another air raid caused serious damage. Two-hundred-and-fifty high-explosive bombs and thousands of incendiary bombs fell on the city and dockyard. The destroyer *Witherington* was hit and beached off the North Corner, and the trawler *Rivello* was sunk. The dockyard power station was hit by a high-explosive bomb and burnt out. Nos 1 and 3 electrical shops were badly damaged by fire. No. 1 Ship Shop and the west end of the Factory were damaged by fire. No. 10 Storehouse was hit by an incendiary bomb, which caused the clock tower and upper floor to collapse. HMS *Victory*, in No. 2 Dock, was damaged by a small bomb under her fore foot. A hole 8ft long and about 15ft in height was torn in the hull planking and its supporting timbers, and a 5ft hole was made in the orlop.[23] The shore establishments HMS *Dryad*, Haslar hospital, HMS *Excellent* and HMS *Vernon* all suffered damage. At HMS *Dolphin*, the floating dock was sunk. Fire at the RN Barracks resulted in the loss of accommodation for 1,000 men. Three tanks at the Forton oil fuel depot were burnt out. Minor damage to a number of warships in the dockyard was also recorded.[24] The main staircase and Victorian ballroom in Admiralty House were damaged by unexploded bombs, and the ballroom was destroyed in a raid on 17 April, when other damage to the building was also sustained. Repairs were underway when further bomb damage occurred on 3 May. The Commander-in-Chief had to move to HMS *Victory* for the rest of the war.[25]

Further damage occurred within the dockyard on the night of 27 April 1941 during a major raid on the city. On the night of 3–4 May 1941, twelve high-explosive bombs hit the dockyard. There was structural damage to No. 6 Boathouse and No 17 Store,

and a bomb hit the cruiser *Sirius*, then fitting out, between X and Y turrets and passed through the ship's side above water. Fires occurred in three of these locations. There was just one serious casualty.[26] The north-west end of St Ann's Church was destroyed, including the cupola and bells. Another victim of bombing was the Royal Navy Academy, which sustained serious damage to the south wing that caused the western wall to lean forward by over 12ft, allowing the cupola to sag. The School of Navigation, then housed there, had to be moved to Southwick House.[27]

Immobilisation in the Event of Invasion

In late 1941, plans were drawn up by the Commander-in-Chief, Portsmouth Command, for the immobilisation of the port and dockyard in the event of invasion by the enemy, and these plans were successively amended over subsequent months through to June 1942. (There may also have been earlier versions of these plans because the invasion threat was at its height in summer 1940.) There were two alternative schemes for stopping enemy vessels entering or berthing in the harbour. One was to block the harbour entrance using two blockships, *Marshal Soult* and Coal Depot C1, which would be sunk (using explosive charges) in the narrows between Tower House and Fort Blockhouse, and for this purpose demolition charges had been fitted to both blockships. The gap between them would be filled by sinking smaller craft there. The alternative scheme was to sink about twenty-five vessels as blockships alongside the berths in the dockyard. These vessels would range in size from the old French battleship *Courbet* to small drifters, and would provide obstructions to the berthing of vessels. HMS *Vernon*, the mine warfare, torpedo and diving establishment at Portsmouth, was to be responsible for the sinking of blockships.[28]

There were complementary plans to immobilise all important facilities and equipment. The caissons of A, B, C and D Locks and No. 9 Dock would be put out of action by removing vital parts. The Oil Fuel Jetty and pipeline would be severed. The Dockyard Generating Station would be immobilised by removing vital parts, and it was noted that the city generating station would also have to be immobilised. The floating cranes, steam cranes, dock pumps, hydraulic pumps, distilling plant, oxygen plant, locomotives, air compressors, diving gear and salvage plant, power boats and telephone exchanges would all be immobilised by removing vital parts or taking similar action. The floating dock was to be sunk and its electric power cut off. Stocks of petrol and oil would all be contaminated. Motor vehicles and yard craft that could not be evacuated would be immobilised. Oilers, dredgers, certain auxiliary craft and tugs would proceed to another area, possibly towing lighters, etc. Slipways, which might be used by amphibious tanks, would be blocked or have their hauling-up gear dislocated. Ships undergoing repair in the dockyard would be evacuated under their own steam, or towage, where possible. The Royal Naval Armament Depots on the Gosport side of the harbour were to be evacuated and immobilised. Secret and confidential books, papers and drawings would be removed to a safe area, buried or destroyed. By the time all these plans were detailed, the threat of invasion had receded, and fortunately they never had to be operationalised.[29]

The cruiser *Achilles* suffered an explosion on board in June 1943 whilst refitting in the dockyard, killing fourteen dockyard men and injuring many others.

Explosion in HMNZS *Achilles*

The cruiser *Achilles*, when part of the Royal New Zealand Navy, spent fourteen months in Portsmouth dockyard refitting and rearming from April 1943 to May 1944. On 22 June 1943, a violent explosion occurred in one of her main fuel tanks, killing fourteen dockyard hands and injuring many others (twelve being sent to hospital), and causing considerable structural damage to the ship. The tank had been emptied and cleaned in April, and workmen were making moulds in the double-bottom fuel tank preparatory to erecting two bulkheads in the compartment. The fuel tank in which the explosion occurred and three other compartments were almost completely wrecked. A number of bulkheads were collapsed or badly distorted by the blast. The deck above was blown upwards about 7ft, the platform deck was torn away from the ship's side and the shell plating bulged outwards over an area of about 30ft x 10ft. A number of watertight doors were blown through their frames.

Besides those killed, a considerable number of dockyardmen, injured or stunned, were trapped in the damaged compartments. They were rescued by members of the ship's company, assisted by other dockyardmen. Dense smoke at first prevented access to the seat of damage. Two ratings equipped with breathing apparatus

tried to get through by way of the stokers' mess deck but were overpowered by smoke, and one had to be hauled out by means of a lifeline. The smoke was finally dispersed by water spray. Some ten or twelve dockyardmen owed their lives to the initiative and cold courage of three ratings who, regardless of their own safety, went below and worked to the limit of endurance. They were Stoker First Class William Dale, RNZNVR, who was subsequently awarded the Albert Medal, and Engine-Room Artificer William Vaughan, RN, and Stoker First Class Ernest Valentine, RNZNVR, who were mentioned in despatches.

Finding that all smoke apparatus was in use by others, Stoker Dale tied a handkerchief over his mouth and made a difficult descent through three decks into a smoke-filled space. The compartment was badly collapsed, but in the darkness its condition was quite unknown to Dale. Without hesitation he got to work and passed up four injured men who were in various stages of collapse. They afterwards affirmed that they could not have got out without help.

Having 'surfaced for a short breather', Dale then went down into the fuel tank in which the explosion had taken place and groped his way in the darkness through debris and thick smoke, and with great difficulty wriggled through the distorted manhole in the tank top. The twisted, vertical steel ladder was far short of the bottom of the tank, but he trusted to luck and landed safely. With equal courage, a dockyard worker named Rogers descended and assisted Dale in rescuing two injured men who were hauled up by ropes.

Wearing a smoke helmet, Vaughan went down into a compartment, the condition of which was unknown, in an attempt to rescue men believed to have been working there. He could not find them in the pitch darkness and dense smoke, and, in an almost unconscious state, had to be assisted back. Recovering after a short spell, Vaughan went down to the switchboard room, from which he sent up several semi-conscious men before he was again almost overcome by fumes and assisted back to the upper deck. On both occasions, Vaughan was saved by the energetic action of a 16-year-old boy named Baxter, who had been boiler-cleaning. Stoker Valentine worked his way through smoke and debris into a badly wrecked compartment and rescued a number of dazed men. It was probably from this or an adjacent compartment that others were rescued later by Stokers Clarke and Stow and Leading Supply Assistant Brittain. Due to the structural damage to the ship, completion of the refit was extended by eight months.[30]

Operation Overlord

Operation Overlord, the Normandy Landings in June 1944, involved what was probably the biggest concentration of military, naval and air power the world has ever seen, and Portsmouth was the hub of the invasion preparations. The Royal Navy had overall responsibility for Operation Neptune, the naval part of the plan. Of the 1,213 warships involved, 200 were American and 892 were British; of the 4,126 landing craft involved, 805 were American and 3,261 were British (the balance in each case were from other Allied countries). By the end of 11 June (D+5), 326,547 troops, 54,186 vehicles and 104,428 tons of supplies had been landed on the beaches. Naval losses for June 1944 included twenty-four warships and thirty-five merchantmen or auxiliaries sunk, and a further 120 vessels damaged. Many of the specialist ships and landing craft used on D-Day had been modified in Portsmouth Dockyard. Parts of the Mulberry Harbours (the artificial harbours that were used by the Allies for landing troops and supplies in Normandy) were built there. It was also an embarkation point for troops.[31]

Operation Overlord was controlled from HMS *Dryad*, the Navigation School at Southwick House, by General Eisenhower, the Supreme Allied Commander. He was assisted by General Montgomery and, as commander of Operation Neptune, Admiral Sir Bertram Ramsay. Stokes Bay, Lee-on-Solent, Hardway (Gosport) and Gosport town all provided major embarkation points for troops into their landing craft.

The vast majority of the British and Empire naval force that landed on Gold, Juno and Sword beaches was controlled by the Portsmouth Command and set out from ports and harbours between Portland and Shoreham. The huge task of maintaining, repairing and modifying the landing ships and landing craft fell on Portsmouth Dockyard. Given that the yard was already working at full stretch, this would have been an impossible task had it not been for the subsidiary landing craft repair bases that were set up in private shipyards or at new purpose-built boatsheds and slipways in nearby locations, usually associated with landing craft bases. The repair bases included HMS *Lizard* at Shoreham, HMS *Sea Serpent* at Birdham, HMS *Northney* on Hayling Island, HMS *Tormentor* on the River Hamble, HMS *Cricket* at Bursledon, HMS *Medina* on Wooton Creek, HMS *Mastodon* on the Beaulieu River, HMS *Turtle* at Poole, HMS *Bee* and HMS *Grasshopper* at Weymouth and HMS *Squid* at Southampton. Maintenance facilities for minor landing craft were provided at Poole on a requisitioned timber merchant's site, as part of HMS *Turtle*, and for landing barges at a facility in Langstone Harbour at HMS *Dragonfly*. Most of the work on tank landing ships (LSTs) was undertaken in Portsmouth Dockyard.[32]

The work in the dockyard included the construction of twenty Phoenix concrete caissons (twelve in C Lock and eight in the floating dock), which were towed to the Normandy beachheads to form part of the two Mulberry Harbours at St Laurent and Arromanches. The Phoenixes were large concrete structures, with hollow chambers inside. They could be floated in order to move them, but when water was let into the internal chambers they would sink onto the sea bottom. Phoenixes were built

Preparations for Operation Overlord in Portsmouth Harbour: tropps aboard a landing craft.

by contractors, those built in the dockyard being by Bovis and Laing (six each in C Lock) and Lind (eight in the floating dock) using skilled labour (such as scaffolders, steel benders and steel erectors) recruited from across the country, plus labourers, many of whom were men past military age supplemented by younger men conscripted into the mining industry and then diverted to the Mulberry project; Irish labourers were also brought in because of a local shortage of able-bodied men. The Phoenixes were also constructed in Langstone Harbour, on the stretch of Hayling Island beach at the harbour entrance. Once built, they were sunk nearby until they were needed. One of the Phoenixes developed a fault and could not be used: it remains there to this day, on a sandbank in Langstone Harbour.[33] Fourteen Phoenixes were locally built at Stokes Bay by some 1,600 men: scaffolders, carpenters,

fitters, steelworkers and Irish labourers working around the clock on shifts, and others at Southampton Docks, Lepe, Marchwood, Stone Point near Fawley and other yards on the Thames Estuary. It was a massive project: in total, 212 caissons were built, each displacing between 2,000 and 6,000 tons, and at its peak some 45,000 workers were employed on it. The workers were often billeted with local families by local government officers who had powers to compel any reluctant hosts.[34]

Units of the Whales – the floating steel roadways and pierheads – and the Bombardons – floating steel breakwaters – were also constructed in the dockyard for use in the Mulberry harbours. Much of the experimental work and some construction of the Pluto pipeline, which was to be laid on the bed of the English Channel between Sandown and Querqueville to supply petrol to

Mary Rose was built at Portsmouth in 151011 and is said to have been King Henry VIII's favourite ship. It remains uncertain whether her loss in 1545 was an accident or the result of damage sustained in action. (Geoff Hunt/The Mary Rose Trust)

HMS *Victory*, Nelson's flagship at the Battle of Trafalgar, is still in commission as a ship of the Royal Navy. She is shown after being repainted in what are believed to be the accurate colours used whilst she was in operational service. Her topmasts and yards have been removed to reduce the weight on her hull during a major conservation programme.

HMS *Warrior* was probably the most significant step in the transition from wood and sail to iron and steam in the Navy. She was launched on 29 December 1860 by Thames Ironworks at Blackwall. Thick frozen snow covered the shipyard and the ship was frozen to the slip, so after the naming ceremony she refused to budge. It took twenty minutes to free her before she slid down into the river.

HM Monitor *M33* was a heavily armed, shallow-draught ship designed to bombard enemy troop positions in the 1915 Gallipoli campaign. Her hull is dazzle painted to make range-finding difficult on enemy ships and submarines.

A model of the Third Rate *Captain* after her rebuild in the dockyard in 1708.

HMS *Queen* was launched at Portsmouth in 1839 and is shown leaving Grand Harbour, Malta. (Robert Strickland Thomas)

The old Royal Naval Academy at Portsmouth was completed in 1733 to board and teach forty young prospective midshipmen, and was the Navy's first shore-based training establishment.

Admiralty House was built as the Dockyard Commissioner's House and is now the residence of the Fleet Commander.

The new Semaphore Tower, Rigging House and Rangefinder was completed in 1930, and incorporated the Lion Gate (seen beneath the tower) which dated from 1778.

The Victorian royal shelter (1893) was originally sited on the South Railway Jetty to accommodate royal family members when transferring from train to royal yacht or ship. It is shown where it is now sited, near the entrance to the dockyard Camber. Berthed in the Camber is Steam Pinnace 199, built at Cowes in 1911 for use aboard battleships.

No. 6 Boathouse (1846) is fronted by the boat pond, formerly a mast pond.

The Mary Rose Museum seen across the stern of HMS Victory. The museum building is elliptical in form, and is said to mirror the shape of the old ship's hull. The steel frame of the building is clad in black-stained red cedar timber in much the same way as the carvel-built hull was constructed. (The Mary Rose Trust/ Mindworks Marketing).

The submarine *Onyx* in No. 12 Dock for 'docking essential defects', 31 August 1988. She was in dockyard hands for twenty-four weeks. (Portsmouth Royal Dockyard Historical Trust)

Looking across C and D locks, 13 July 2005. A Type 23 frigate is berthed on the North-West Wall, whilst in D Lock is another of the same class and C Lock (right) holds the ice patrol ship *Endurance*. In the pocket (left) are a Type 42 destroyer and a Sandown-class minehunter.

A view across the dockyard during the 2005 International Festival of the Sea.

The new aircraft carrier *Queen Elizabeth* berths at Portsmouth for the first time, 16 August 2017.

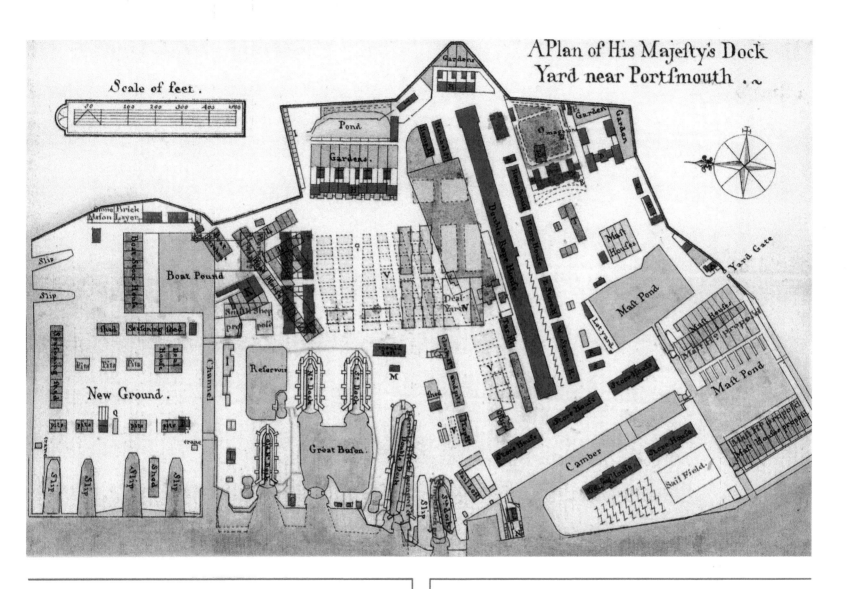

A plan of the dockyard, 1786. Although much of the work under the 1761 Plan was complete by then, this drawing does show some further proposed changes superimposed on the existing layout.

A sheer hulk in Portsmouth Harbour with a ship alongside on which the masts are being stepped, by Edward Cooke, 1836.

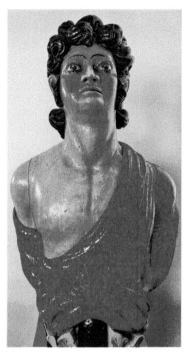

The figurehead of HMS *Orestes* displayed in the NMRN at Portsmouth. The ship was an eighteen-gun sloop launched at Portsmouth in 1824, which in 1852 became a coal hulk at Portsmouth and was finally broken up in 1905. In Greek legend, Orestes was a son of Agamemnon, a Greek king.

The figurehead of HMS *Actaeon*, displayed in the NMRN at Portsmouth. The ship was a twenty-eight-gun Sixth Rate frigate launched at Portsmouth in 1831 which took part in the Second China War in 1857, and was hulked in 1870. *Actaeon* was a hunter whom the goddess Artemis turned into a stag. He was then torn to pieces by his hounds.

The figurehead of HMS *Eurydice*, displayed in the NMRN at Portsmouth. The ship was a twenty-six-gun Sixth Rate frigate launched at Portsmouth in 1843. Used as a training ship for boys, she sank with the loss of 318 men and boys (almost all her crew) off the Isle of Wight on 24 March 1878. She was raised six months later and taken to Portsmouth for breaking up. Eurydice was the wife of the Greek hero Orpheus. Her story is very sad, which is why the figurehead has such a grief-stricken expression.

The figurehead of HMS *Bellerophon*, displayed in the NMRN at Portsmouth. The ship was a Third Rate battleship launched at Portsmouth as HMS *Waterloo* in 1818, but was renamed *Bellerophon* in 1824. During the Crimean War, she took part in the bombardment of Sebastopol in 1854, and became a harbour training ship at Portsmouth in 1856 before finally being broken up in 1892. Bellerophon was a Greek warrior who killed the monster Chimaera with the help of the winged horse Pegasus.

The Porter's Garden was restored between 2000 and 2003, using the principles of eighteenth-century garden design and contemporary maps of the garden, and in 2009 a raised extension was opened in front of No. 6 Boathouse. The garden is outlined by the dockyard wall, a pleached lime hedge, yew and box hedges and four walnut trees. Local schoolchildren designed the knot garden.

No. 5 Boathouse was formerly a mast-making house which was rebuilt in 1882. Its wooden construction echoes the appearance of the many wooden dockyard buildings erected in the seventeenth century before brick became the common construction material.

HMS *Warrior* and the dockyard seen from Spinnaker Tower in November 2007.

The Tidal Basin in April 2015, showing the entrance to C Lock on the left and the destroyer *Defender* berthed on the right.

The launch of the 3rd Rate *Warrior* at Portsmouth on 18 October 1781, depicted in a model in the National Museum of the Royal Navy, Portsmouth.

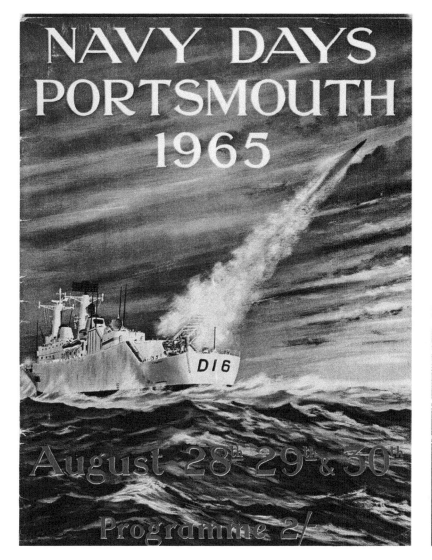

The cover of the programme for 1965 Navy Days, showing the guided missile destroyer *London*, which was open to the public, as were the aircraft carrier *Centaur*, the cruiser *Lion*, five other destroyers and frigates, four submarines, two minesweepers and RFA *Tideflow*. There were many events and displays, including two landing craft (LCA) offering harbour trips from King's Stairs, flypasts by Fleet Air Arm aircraft and the submarine *Token* conducted a diving display each day in No. 2 Basin.

The minehunter *Iveston* in No. 8 Dock, with No. 2 Basin and the Steam Factory beyond. (Portsmouth Royal Dockyard Historical Trust)

The launch board for the frigate HMS *Sirius*. It was traditional for the painters to produce such a board to be displayed on the launching platform when a ship was launched at Portsmouth. Sirius was the god or goddess of the Dog-Star, the brightest star of the constellation Canis Major. The pre-dawn rising of the star in the path of the sun was believed to be the source of the scorching heat and droughts of midsummer, and marked the flooding of the Nile in Ancient Egypt and the 'dog days' of summer for the ancient Greeks.

SUCCESS TO H.M.S. SIRIUS.

The launch board for the frigate HMS *Andromeda*, the last ship to be built in the royal dockyard. In Greek mythology, Andromeda was the daughter of the Aethiopian king Cepheus and his wife Cassiopeia. When Cassiopeia's hubris leads her to boast that Andromeda is more beautiful than the Nereids, Poseidon sends the sea monster Cetus to ravage the North African coast of Aethiopia as divine punishment. Andromeda is stripped and chained naked to a rock as a sacrifice to sate the monster, but is saved from death by Perseus.

SUCCESS TO H.M.S. ANDROMEDA.

the advancing troops in Normandy, were undertaken in the yard. The first part of the pipeline was laid by the Hopper Barge W24 (renamed *Persephone*) which was converted in the dockyard.[35]

Operation Overlord called for much innovation, and the dockyard was involved in leading many development projects. Nineteen tank landing craft (LCT) were converted to LCT(R) by the installation of Army type rocket projectors, for use in bombarding enemy troops near the beaches. Large numbers of LCT were fitted with ramp extensions for launching Sherman tanks, and others were armoured as LCT(A) to provide close-in gunfire support in the early stages of the landings. Eighteen LCA were converted with hedgerows, forward throwing mortars. Floating mat bridges, known as Swiss Rolls, were developed in the yard and the prototypes were manufactured there.[36]

In the period between April 1944 and the sailing for the assault on 5–6 June, the work in the dockyard increased considerably, the numbers of ships dealt with by the dockyard for refits, boiler cleaning, defects, changing guns, etc, being as follows:[37]

Battleships, cruiser, monitor, destroyers, fleet minesweepers	88
LST	16
LCT	186
Trawlers, drifters, boom defence vessels	30
Miscellaneous craft, oilers, dredgers, yard craft, etc.	53
Coastal forces craft	33

In addition to the above, twelve frigates and Hunt-class destroyers were fitted with bow chasers, twenty-four minesweepers were fitted with new equipment, HM ships *Durban*, *Centurion* and HMNS *Sumatra* were prepared as blockships and HM ships *Despatch*, *Largs*, *Hilary* and *Bulolo* were fitted out as headquarters ships for force commanders, with the addition of much radio and radar equipment.

For a few days after D Day, the dockyard was very little involved in the operations from a repair point of view and was mainly concerned with the supply of stores of all descriptions and the evacuation of casualties returning from Normandy. Four days after D Day, ships began to return for repairs and the dockyard again became heavily involved. For example, fifty-three mechanised landing craft (LCM) were in hand for repairs by 17 June. During the

period between 6 June and 31 July alone, work on no fewer than 513 vessels was undertaken by the dockyard or local firms under its general control, as follows:

	In hand	Dry-docked	Completed
Destroyers, fleet m/s, and larger	76	15	62
LST	44	22	41
LCT	247	203	197
Trawlers, drifters, etc.	65	16	39
Miscellaneous craft	25	13	24
Coastal forces craft	56	33	55

Large damage repairs to *Rattlesnake*, *Wrestler*, *La Surprise*, *Halstead*, *Pink*, *Persian*, *Kellett*, *Trollope*, USS *Nelson*, LST 359, *Fury*, *Goathland* and *Apollo* were patched and the vessels removed elsewhere for complete repairs.[38]

Battleships that were part of the Normandy bombarding force returned to Portsmouth during Operation Overlord, but did not receive much dockyard attention. The monitor *Roberts*, also part of the bombarding force, underwent repairs in the yard to one of her 15in guns which had a burst jacket following arduous service off Normandy, and the cruiser *Arethusa* was under repair for mine damage.[39]

Working in the Dockyard During the War

The war brought special pressures to bear on those working in the yard, and there was always the threat of air raids. The big ships were less in evidence. 'There were mostly small ships as it was too risky for big ships, although I do recollect HMS *Resolution* coming in and being repaired by planking, by gangs of shipwrights, after being torpedoed before it went to America for repair,' said one retired shipwright. There were vivid recollections of the wartime scene. '[S]treams of exhausted soldiers returning from the Dieppe Raid and the stables being used as a makeshift mortuary,' recalled another retired shipwright, who also remembered the day a landmine fell 'on to one of the 250-ton cranes', with the parachute remaining suspended whilst the mine fell onto the capstan machinery below. 'Luckily it didn't go off as there was the cruiser *Berwick* on one side and the liner *Vienna* the other.'[40]

Former electrical fitter apprentice D.K. Muir recalled that the switch from daylight to night air raids made attendance at Dockyard School evening classes even more onerous than before. On Friday, 10 January 1941, an air raid sent the class down to the basement of the college, crouched under tables for added protection. Muir said:

> The lecturer with typical British phlegm attempted to continue the lesson as if nothing was happening but the scream and explosion of bombs and the noise of gunfire made it clear that this was no ordinary raid and we remained in the basement until 11 p.m.

He emerged during a lull in the noise to find the 'Town Hall' ablaze and the railway station, his only means of returning to his home in Emsworth, bombed and out of commission. There was nothing for it but to walk the 12 miles home, which he did, getting there at 3 a.m.[41]

Working conditions in the dockyard were still pretty basic: there were no rest rooms or facilities for changing or washing clothes. Workers would often build their own shelters, or 'cabooses' as they were called. One retired shipwright remembered:

> People don't realise that things were very hard in those days. When I went into No. 3 Shop, in about 1943 ... do you know there wasn't a tap in the shop ... no toilets, no washing facilities, and not even a drinking tap, the only taps were for the drills, keeping the machinery cool, you had to walk outside the shop to find a cold water tap ... you had to walk across the road to use the toilet and they were just long rows with no doors, no seats ... they were slabs of slate with running water underneath so in the winter if that froze you can imagine what it was like! That was the conditions up to the end of the war when they put wash basins and toilets in No. 3 Shop ... The conditions were shocking, but it was a job in those days.[42]

The north-east corner of No. 3 Basin, 25 May 1944. The three submarines on the right foreground are *Tireless* and *Token*, which were fitting out, and outboard of them the incomplete French *Ondine*, which had been towed to Portsmouth in 1940 just after launching to prevent her falling into German hands. In the background is the floating dock AFD 11, and top right is the dockyard Round Tower. This tower had been built in 1843–48 and was moved to its present position in 1867. (Portsmouth Royal Dockyard Historical Trust)

The influx of women in 1941 was to be a mixed experience for them. A former red lead painter recalled:

I had to work during the war. If our children were evacuated we had to work … they find everything out, don't they? They sent you to different places, my sister-in-law worked in the machine shop … I started with £2 a week, which was good money.

A temporary clerical assistant recalled that she too had to work:

Broke my heart, but I wanted the money. I had the mortgage to pay on the house and I lost my husband in the first three months of the war, so I didn't have it easy so I had to find a job … I started at £3 10s a week and felt ever so rich.

Some of the women were overawed by the dockyard environment. A former electrical fitter said:

My first impression … it was a vast place and there were so many people milling about, like little ants really, they were dashing everywhere and they knew where they were going and I never knew anything! I felt so stupid really, because this was something I'd never done in my life, never mixed with so many people and never seen such a huge building, it was very scary. Past the gate, past the police, and that's when you got the importance of it, and you knew that this was a very important place.

But she adapted to this new world and became involved in preparing HMS *Hilary*, a command ship for the D-Day invasion:

[T]here was this air of hustle and bustle and importance and 'It's got to be done for the invasion' and everything. It was a happy, happy time and they made it as comfortable as they could for everybody, that was how I viewed it … once I was on that ship I used to think 'Oh this is wonderful, you're doing something that nobody else is doing, I'm doing my bit for the war effort!'[43]

The women worked as wire splicers, tracers in the drawing office, machinists, stud welders, acetylene welders, riveters, riggers, laggers, colour-makers, electrical fitters and clerical assistants. They received less pay than the men. The electrical fitter added:

It wasn't discussed to the extent it is now, it was accepted that that was the working rules, that you didn't get the same as the men and nobody hit out about it. These men, when we were called up to do the job, you couldn't possibly have all the knowledge that those men had who had a lifetime doing it, they had had apprenticeships, they'd been in the dockyard for years, they knew all the ropes … you were just there for wartime, so you couldn't expect to get their money even if you'd have thought about it … The women couldn't do the fitting job, we were only labourers.

One woman, asked how long it took to learn the skill of wire splicing, replied, 'I suppose about a week. You couldn't work with gloves and where you'd have the wire on your knee, where you were splicing, so they were black and blue 'till you got used to it.' Another who learned to do the lagging said:

'This lag down this post,' I says to the chargehand alongside of me, 'Not much of a job is it?' So he says, 'Can you do better?' I said, 'Well, I think so.' So I had to do this, so he says to me, 'Oh you've got the job.' I says, 'Any increase in money?' He says, 'No' … Well at the end of the week they told me … I would do all the lagging. I used to do boiler tops you know and all that.[44]

16

THE COLD WAR

The Cold War Era

As the nation thankfully returned to peace, hundreds of British warships were paid off, their crews were demobbed and a slimmed-down Navy emerged, consisting mainly of the most modern of the war-built ships. As well as the active fleet, large numbers of ships were kept in reserve because a new threat was emerging from the Soviet Union, which was building a huge submarine fleet (with over 300 boats by 1949), and the Royal Navy instituted a programme of converting war-built destroyers into fast anti-submarine frigates. In response to the Soviet threat, the North Atlantic Treaty Organisation (NATO) was formed in 1948, creating an alliance between the USA, United Kingdom, Canada and the western European countries. The Cold War that ensued for over forty years would shape the role of the Royal Navy and, for much of that time, keep the main dockyards in business.

The outbreak of the Korean War in 1950 gave impetus to moves to reconstruct the Navy with more modern warships, and orders were placed between 1950 and 1952 for new frigates (including one from Portsmouth Dockyard), submarines, minesweepers, seaward defence boats and fast patrol boats; work on the construction of four aircraft carriers and three cruisers was resumed in private yards after having been suspended at the end of the Second World War.[1] In 1949–50, 15,400 people were employed in the Portsmouth yard; in 1950–51, the demands of the Korean War increased the number to 16,500. Even after the war had ended, the surge in numbers continued: by 1954 it had risen to 17,472, and following a drop of about 1,000 it went up again to 17,480 in 1958. Remarkably, this

figure was actually two more than the peak number seen in the Second World War, in 1943. By 1963, employment had declined somewhat, but was 'well over 16,000'.[2] However, the post-war era saw gradual but dramatic reductions in the strength of the Navy that would inevitably lead to downsizing in the dockyards. The number of RN personnel declined from 167,000 in 1948–49 to 121,500 in 1957, 77,467 in December 1968 and 59,300 in 1979.[3] Concurrently, the size of the fleet was slashed: in 1952, there were thirteen aircraft carriers, five battleships, twenty-six cruisers, 271 destroyers and frigates and fifty-three submarines; by 1982 there were only two aircraft carriers, two assault ships, fifty-nine destroyers and frigates, thirty-one submarines and no battleships or cruisers.[4] As a result, the number employed in Portsmouth Dockyard had fallen to 6,925 by 1981.[5]

In the immediate years after the war, the dockyard was concerned with refitting ships that had returned from war service and would be retained in the post-war fleet, de-equipping ships that were to be scrapped and 'mothballing' ships that were to be retained in the reserve fleet. However, in 1949 a rearmament programme was started and a number of war-built ships were scheduled for modernisation. Thus, in the same year, the destroyer *Relentless* entered the yard for conversion to an anti-submarine frigate, which was completed in July 1951. Her superstructure was stripped down to upper-deck level and rebuilt with two Limbo anti-submarine mortars, a new sonar outfit and a new gun armament.[6] Together with *Rocket* (converted at Devonport), she was one of the first pair of such conversions. A further twenty-one destroyers were converted, including two at Portsmouth – *Verulam* (1951–52) and *Troubridge* (taken in hand in 1955 for the structural work, though the final fitting out was completed by J. Samuel White at Cowes and completed in 1957).[7]

After the war, it was decided that the Illustrious-class aircraft carriers would have to be reconstructed to be able to operate modern aircraft. An increased hangar height of 17ft was needed and aircraft weight would be up to 30,000lb, compared with 20,000lb, at most, in the original design. *Victorious* was selected to be reconstructed first and was taken in hand at Portsmouth

The fast anti-submarine frigate *Verulam* in 1952, newly converted from a fleet destroyer in the dockyard.

The aircraft carrier *Victorious* after her major modernisation in the dockyard, which lasted seven years. She is seen leaving Portsmouth Harbour.

in October 1950 for what turned out to be a very lengthy and expensive modernisation, with the ship being rebuilt from the hangar deck up. The intention was to complete the work by April 1954 (at a cost of £5.4 million), but the reconstruction was delayed by changes to the specification and the belated decision that the boilers would have to be replaced, by which time the armoured deck had been refitted, resulting in a considerable amount of completed work being dismantled. A new 8½°-angled flight deck, two steam catapults, Type 984 radar and six twin 3in guns were fitted, and her full load displacement increased from 28,600 to 35,500 tons. She took to the water again on 12 December 1955 after four and a half years on the blocks in D Lock, and work was finally completed in January 1958, at a cost of £30 million.[8] The delays to *Victorious* and the escalation of the costs involved led to the decision to carry out no further Illustrious-class modernisations.

Between 1950 and 1959, 2,047 vessels were dry-docked or slipped, and the period of the Suez Crisis in 1956 was a particularly busy time.[9] No new construction was under way, until the frigate *Leopard* (ordered on 21 August 1951) was laid down on 25 March 1953, when Mrs A.G.V. Hubback, wife of the Admiral Superintendent, switched on an automatic welding set that joined together two sections in the then modern equivalent of the traditional keel-laying ceremony. The ship was launched by Princess Marie Louise on 23 May 1955, cheered on by a crowd of nearly 10,000 people. Her bow was built in No. 1 Ship Shop and she was the first prefabricated, all welded warship to be built in the yard.[10] In 1955, 340 apprentices were recruited into the yard, though in the next few years this declined, 254 being entered in 1959.[11]

Leopard was completed in September 1958, and a further four frigates were built: *Rhyl* (1958–60), *Nubian* (1959–62), *Sirius* (1963–66) and *Andromeda* (1966–68). Two floating docks for nuclear submarines were built, AFD 59 in No. 15 Dock in 1959–60 (launched 31 March 1960), for service at Barrow, and AFD 60 in C Lock in 1964–66, for docking Polaris submarines at Faslane. The completion of *Andromeda* in December 1968 marked the end of an era for, although it was not known at the time, she was to be the last ship built at the royal dockyard after over 450 years of shipbuilding.

Even during this period of frigate construction, the yard relied on the large modernisations and conversions of ships to supplement the

routine refits and repairs. The aircraft carrier *Triumph* was taken in hand in December 1957 for conversion to a heavy repair ship, and with suspensions to the work caused by other priorities the work was not completed until January 1965. Meanwhile, between May 1958 and 1962, the depot ship *Maidstone* was modernised and adapted to serve nuclear submarines. The aircraft carrier *Bulwark* was in hand from November 1958 to January 1960 for conversion to a commando carrier, and her sister ship *Albion* followed from January 1961 to August 1962. Between 1959 and 1961, the maintenance ship *Mull of Kintyre* was modernised and adapted to support minesweepers. The destroyer *Agincourt* underwent a conversion to serve as a fleet radar picket between December 1959 and May 1962.[12]

The cruiser *Blake* was given a major conversion to a command helicopter cruiser, on which work started in April 1965. The work was delayed by a fire which broke out on an amidships mess deck in July 1966, and it took the Dockyard and City Fire Brigades over two hours to get the blaze under control. Damage to the cruiser's after section, where the 6in turret was being removed, was described as 'fairly extensive'. In January 1969, with the ship dry-docked in C Lock and the work almost complete, another fire – this time in a storage area – broke out. The lock's sluice gates were opened by *Blake's* fire and emergency parties to let water in so that the ship's fire-main system could be made available. Once again the Dockyard and City fire appliances were in attendance and it took two hours to extinguish the blaze. This delayed completion by two months and the ship was finally recommissioned in April 1969.[13]

In May 1966, the frigate *Yarmouth* was taken in hand to begin a long refit equipping her with the Seacat guided missile system and a hangar and flight deck for a Wasp helicopter, the work being completed in September 1968. Between 1960 and 1969, 1,990 vessels of all types were dry-docked or slipped, only a slight reduction on the previous decade.[14]

In 1968, the government decided to withdraw forces from east of Suez, which would lead to the withdrawal of fixed-wing aircraft carriers and a reduction in the size of the fleet. One of the consequences was the 1969 Defence Review, which announced that the royal dockyards would reduce their labour force by 5,000 by the mid 1970s. The first redundancies at Portsmouth were announced in March 1972, with 137 jobs lost.[15]

In January 1970, work began on converting the destroyer *Matapan* to a sonar trials ship. She was rebuilt from the weather deck up and had a new bulbous bow and a large underwater sonar dome fitted whilst in No 14 Dock, where she was docked on 13ft-high blocks and plinths to facilitate the construction of a deep skeg on the bottom of the ship, and the work was completed in February 1973.[16] Between 1973 and 1977, the frigate *Arethusa* was modernised, with the Ikara anti-submarine missile system replacing her twin 4.5in gun turret. However, despite a few large projects such as these, the 1970s was a period of retrenchment in the dockyard as the workload gradually declined. Between 1970 and 1979, 1,042 vessels of all types were dry-docked or slipped, a big reduction on the previous decade.[17] It was decided that in most cases it was more economical to build new ships than carry out expensive midlife modernisation of ships, and this hit the yard badly. By 1978, the workforce had reduced to 8,325.

After the war, Navy Days had been introduced as a replacement for Navy Week; the dockyard was open to the public for four days each year, but this was soon reduced to three. A typical example of post-war Navy Days was that of 5–7 August 1961. The line-up of ships open to the public included the aircraft carrier *Hermes*, the new cruiser *Blake*, destroyers *Trafalgar*, *Dunkirk* and *Carron*, frigates *Puma*, *Berwick*, *Eastbourne*, *Keppel*, *Volage* (from reserve) and New Zealand's *Taranaki*, and submarines *Turpin*, *Tireless*, *Trenchant and Tally Ho!*, as well as smaller vessels and the RFA *Black Ranger*. Amongst the many other attractions were an air display by six Scimitar aircraft of 800 Squadron, a helicopter search and rescue display, two torpedo firings by *Trafalgar*, displays by Royal Marines Gemini craft and Special Boat Section, a diving display by divers from HMS *Vernon*, a seaboat display from the minelayer *Plover*, trips in landing craft, demonstrations of atomic defence and damage control methods, an exhibition of dockyard apprentices' work at No. 3 Shipbuilding Shop (including the Constructive, Marine Engineering, Electrical Engineering and Navy Works departments) and music from the bands of the Royal Marines.[18]

The dockyard steam tug *Capable* was used for berthing and ship towage duties in the harbour and Spithead. A very powerful vessel, she was also capable of salvage and sea towing duties and was sometimes called out into the Channel for salvage work. She was completed in 1946 and served at Portsmouth from then until 1971.

Dockyard Facilities

Dockyard tugs were part of the Admiralty Yard Craft Service and were civilian-manned, under the control of the Captain of Dockyard. In 1959, the vessels were reorganised into the Port Auxiliary Service (PAS), together with the yard craft of the victualling, stores and ordnance departments, in the interests of efficiency and economy. This arrangement lasted until 1 October 1976, when the Royal Maritime Auxiliary Service was formed and absorbed the PAS craft, as well as other vessels in government service.[19] In 1996, the private operator Serco Denholm Ltd (now known as Serco Marine Services, part of Serco Group plc) was contracted to provide the management, operation and maintenance of tugs and other yard craft following a commercial tendering process.

The large floating dock AFD 11 remained at Portsmouth until it was sold in 1959 and replaced by AFD 26, which was moved from Portland. It stayed until 1984, during which time it docked 400 submarines and seventeen other vessels, before being relocated to Rosyth.[20] There was no major modernisation programme for dockyard facilities in the post-war era. Instead, various piecemeal upgrades and a few new buildings were provided, with little sense that the longer-term future of the yard was being assured. In 1955, St Ann's Church, which had lost its west end through wartime bombing, was reconstructed from the original 1785 drawings.[21] In 1960, the former Great Ropehouse, which at 1,095ft was the longest building in Europe, was divided in two, gutted, restored and had the roof replaced, and it continued in use as No. 18 Storehouse.[22] In 1961, the Fleet Headquarters was built, a

four-storey red brick structure whose west elevation, facing HMS *Victory*, has a neo-Georgian style with forty-four windows.[23] In 1962, Fountain Lake jetty was rebuilt.[24]

In the 1960s, an order was issued for the demolition of the three Great Storehouses, for which the Navy had no further use, to provide car parking space. Fortunately, the civic authorities intervened and a campaign secured agreement from the Navy that the buildings should be preserved. As a result, the Royal Naval Museum was established in 1972 on the ground floor of No. 11 Storehouse, incorporating the former Victory Museum and a collection of Nelson memorabilia donated by Mrs Lily McCarthy, a wealthy American whose family had made their fortune from Listerine mouthwash. Mrs McCarthy had also supported the campaign to save the Great Storehouses. When the Navy announced that the aircraft carrier *Ark Royal* would fetch her collection from New York, she objected, saying that Nelson would have not have known what a carrier was. In response, the frigate *Lowestoft*, namesake of the ship in which Nelson had been a lieutenant, was diverted from Hong Kong to the United States.[25]

On 8 February 1969, the Block Mills closed after over 150 years of production: the much-reduced manufacture of blocks was transferred to the first floor of No. 6 Boathouse.[26] In 1977, the dockyard rail link from Portsmouth Harbour station closed.[27] This had connected the South Railway Jetty to the main line and was used by royal trains, the last of which was on 6 May 1939, when King George V and Queen Elizabeth had boarded the liner *Empress of Australia* for an official visit to Canada and the USA.[28] There was a minor land extension to the yard when a new Unicorn Gate was built southwards along Unicorn Road, and the original Unicorn Gate (1779) was isolated as a monument on a roundabout.[29] Originally a gateway through Unicorn Ravelin in the Portsea fortifications, it had previously been moved and re-erected within the new 1865 dockyard perimeter wall.[30] In 1970, the Dockyard Technical College closed.[31] In the 1970s, the area of the dockyard was extended when parts of seven streets (Copenhagen, Abercrombie, Nile, Trafalgar, Duncan, Conway and Chalton Streets), which had been largely cleared of housing, were taken into the yard, forming a wedge-shaped plot between Flathouse Road and Market Way.[32]

The dockyard railway closed in 1977.[33] In 1952, there had been 27 miles of railway within the dockyard, covering both the Georgian and Victorian areas. The viaduct line into the South Railway Jetty was damaged by bombing in the Second World War and dismantled after the war. The last steam loco had been retired in 1962, whilst diesel locos continued. One diesel engine, built in 1914 for the dockyard Engineering Department, had worked until about 1960. Three steam locos were retained as mobile dockside boilers, but were finally sold in 1972.[34]

The old Boiler Shop closed on 3 March 1973, there being less work for boilermakers as gas turbines replaced steam turbines in surface warship propulsion, and in the following month the Sawmills and old Saw Sharpening Sheds were closed. A new Combined Workshop (or Heavy Plate Shop, later known as the Steel Production Hall and in the 1990s as the Constructive Workshop) was built on the site of the Boiler Shop and opened on 11 February 1976. Amongst the trades that relocated to the new facility were the boilermakers, shipwrights, smiths, welders, caulkers, locksmiths and painters.[35] No. 3 Ship Shop, the oldest surviving iron arched building, which had been built in 1844 as a cover to the building Slips Nos 3 and 4, and had now been vacated by the shipwrights, was demolished in 1980. In 1977. the cranes at the Main (No. 5) Slip were dismantled in preparation for the filling in of Nos 1, 2 and 5 Slips and the redevelopment of the North Corner.[36] This began in 1979, combining North, Middle and South Slip Jetties into one linear jetty, known as Middle Slip Jetty, with Sheer Jetty to the south.[37] On 26 May 1978, Haslar Gunboat Yard, built in 1856–58 and an outstation of the dockyard where about ninety men still worked, closed.[38]

Portsmouth Dockyard was not in receipt of any major post-war modernisation project, whilst Rosyth (mid '60s), Chatham (1968) and Devonport (1981) all received nuclear submarine refitting facilities, and at Devonport a covered frigate refitting complex was completed in 1977. This did not augur well for the future of Portsmouth.

Opposite: The submarine *Alaric* in No. 9 Dock, August 1968.

17 THE RUNDOWN OF THE DOCKYARD

The date 25 June 1981 was a black day for Portsmouth as John Nott, Secretary of State for Defence, announced the results of a Defence Review to the House of Commons. Portsmouth Dockyard was to be run down to a Fleet Maintenance and Repair Organisation with the loss of over 5,700 jobs out of the 6,925 employed on that date. It was a body blow to the city – only 1,200 jobs would remain.

In future, the yard would concentrate on the support and repair of HM Ships in operational time during Assisted Maintenance Periods, Self-Maintenance Periods, Docking and Essential Defects Periods and alongside when operating. Major work such as reconstructions and modernisations would no longer take place at Portsmouth, though some refits would still take place there. The large midlife modernisations of ships that had been such a feature of post-war dockyard work were being discontinued by the Navy, and most of the refit work would be undertaken at Devonport and Rosyth. The majority of HM Ships and RFAs for disposal would be de-equipped and prepared for disposal at Portsmouth. There would be a substantial increase in the quantity of stores held within the Portsmouth base as a result of the closure of the RN store depots at Woolston and Llangennech and the vacation of Thorney Island. Redundant dockyard buildings would accommodate the additional storage.[1]

On 9 September, John Nott visited the yard and attempted to address a crowd of 2,000 employees outside the Central Office Block to fulfil a promise he had made to the trades unions at national level when the cuts were announced. He was booed and jeered at, and met with a barrage of iron bolts, hammers, tomatoes and eggs. Glass panels in the entrance and doors behind him were broken, and the minister had to retreat back inside the building and was smuggled out from a rear exit.[2] As he made his way to a coach, he was reported to have been punched in the face.[3]

In March 1982, the number of jobs to be retained was increased to 1,300. On 2 April 1982, the first redundancy notices were issued, but ironically on the very same day Argentina invaded the Falkland Islands and unlimited overtime was immediately instituted to prepare the aircraft carriers *Hermes* and *Invincible* for sailing to the South Atlantic. Further redundancy notices were suspended and those already issued were withdrawn.[4]

This gave the dockyard a temporary reprieve as the urgent work of preparing ships for the Falklands Task Force started. Demarcation barriers were set aside and everyone in the yard worked long hours, working hard in often intolerable conditions. The assault ship *Intrepid* had been paid off and was being de-stored for possible sale, whilst the *Hermes* was undergoing a dockyard-assisted maintenance period and *Invincible* was having operational defects put right whilst her crew were on Easter leave. That weekend, aircraft were flown aboard *Hermes* and *Invincible* as they were stored, loaded with armaments and fuelled, and the essential maintenance tasks were completed. On Monday, 5 April, just three days after the decision to send a task force, they sailed from the port for the South Atlantic. The commanding officer of *Hermes*, Captain Lyn Middleton RN, signalled to the dockyard, 'There is only one way to describe the support we have had from you – MAGNIFICENT', and many other ships sent similar signals of praise. Other ships prepared at Portsmouth for the task force included the assault ship *Fearless* (as well as her sister ship *Intrepid*), the destroyer *Bristol*, the frigate *Diomede*, the survey ships *Hydra* and *Herald* – which were converted into hospital ships – the patrol ships *Leeds Castle* and *Dumbarton Castle* and five Royal Fleet Auxiliaries, one of which, *Stromness*, had been paid off for disposal. Work was brought forward to prepare other ships for further Falklands operations, including the new carrier *Illustrious* and the destroyers *Fife*, *Newcastle*, *Liverpool*, *Southampton* and *Birmingham*. Eighteen merchant ships that were taken up by the Navy for support roles were also prepared at the dockyard for their new duties. After the conflict, damage repairs were undertaken on

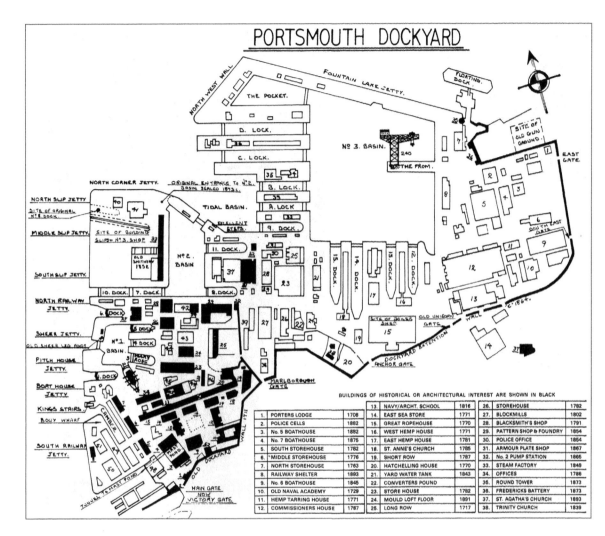

PORTSMOUTH DOCKYARD

No.	BUILDING TITLES	DATE
1.	STORE	1959
2.	No. 52 STORE (TIMBER)	1973
3.	TIMBER STORE	1955
4.	OLD COPPERSMITH'S SHOP	1924
5.	AMALGAMATED PIPE SHOP	1974
6.	STORE	1950
7.	SMALLCRAFT WORKSHOPS	1981
8.	BOX SHED & DINING ROOMS	1889
9.	No. 1 ELECTRICAL SHOP	1945
10.	OLD TORPEDO & PARAVANE STORE	1917
11.	TOOL ROOM	1944
12.	THE FACTORY	1906
13.	OLD TORPEDO WORKSHOPS	1899
14.	APPRENTICE TRAINING CENTRE	1980
15.	HEAVY PLATE SHOP	1975
16.	12 & 13 DOCK COMPLEX	1979
17.	EAST OFFICE BLOCK	1982
18.	14 & 15 DOCK COMPLEX	1979
19.	WOOD STORE	1945
20.	NAVAL DETENTION QUARTERS	1911
21.	WEAPONS ELECTRICAL WORKSHOP	1936
22.	MACHINE SHOP	1945
23.	GUNNERY EQUIPMENT SHOP	1896
24.	MACHINE SHOP	
25.	No. 1 PUMPING STATION	1870's
26.	OLD ELECTRICAL GENERATOR STATION	1907
27.	No. 2 ELECTRICAL WORKSHOP	1954
28.	STORE (OLD SMITHERY)	1896
29.	LAGGING CENTRE	1975
30.	PAINT SHOP	1896
31.	GALVANIZING SHOP	1905
32.	BOX SHED & STORE	1972
33.	BATTERY SHED	1915
34.	No. 4 PUMPING STATION (NORTH)	1911
35.	RIGGING SHEDS	1969
36.	OFFICE & BOX SHED COMPLEX	1975
37.	LIGHT PLATE SHOP (EXTENTION)	1939
38.	STORE	1959
39.	(C.O.B.) CENTRAL OFFICE BLOCK 1 & 2	1966 1973
40.	OLD WELDING SHOP	
41.	OLD SMITHERY. NOW F.M.U.	
42.	STEEL FOUNDRY	1926
43.	JOINERS SHOP	1912
44.	SURGERY & STORES	1902 1894
45.	OFFICE BLOCK	1960
46.	TELEPHONE BUILDING	1903
47.	DOCKYARD CENTRAL LABORATORY	1848
48.	No. 4 BOATHOUSE	1939
49.	SEMAPHORE TOWER	1926
50.	CHAIN TEST HOUSE	1905
51.	STORE	1905
52.	FILM CORP. OFFICE	c.1800

BUILDINGS OF HISTORICAL OR ARCHITECTURAL INTEREST ARE SHOWN IN BLACK

1.	PORTERS LODGE	1708	13. NAVY/ARCHT. SCHOOL	1816	26. STOREHOUSE 1782
2.	POLICE CELLS	1882	14. EAST SEA STORE	1771	27. BLOCKMILLS 1802
3.	No. 5 BOATHOUSE	1882	15. GREAT ROPEHOUSE	1770	28. BLACKSMITH'S SHOP 1791
4.	No. 7 BOATHOUSE	1875	16. WEST HEMP HOUSE	1771	29. PATTERN SHOP & FOUNDRY 1854
5.	SOUTH STOREHOUSE	1782	17. EAST HEMP HOUSE	1781	30. POLICE OFFICE 1854
6.	MIDDLE STOREHOUSE	1776	18. ST. ANNE'S CHURCH	1785	31. ARMOUR PLATE SHOP 1867
7.	NORTH STOREHOUSE	1763	19. SHORT ROW	1787	32. No. 2 PUMP STATION 1865
8.	RAILWAY SHELTER	1893	20. HATCHELLING HOUSE	1770	33. STEAM FACTORY 1849
9.	No. 6 BOATHOUSE	1845	21. YARD WATER TANK	1843	34. OFFICES 1786
10.	OLD NAVAL ACADEMY	1729	22. CONVERTERS POUND		35. ROUND TOWER 1873
11.	HEMP TARRING HOUSE	1771	23. STORE HOUSE	1782	36. FREDERICKS BATTERY 1873
12.	COMMISSIONERS HOUSE	1787	24. MOULD LOFT FLOOR	1891	37. ST. AGATHA'S CHURCH 1893
			25. LONG ROW	1717	38. TRINITY CHURCH 1839

A plan of the dockyard showing the layout c.1980. (B. Patterson/PRDHT)

the destroyers *Glamorgan*, *Antrim* and *Glasgow*, and the submarine *Onyx*, which had suffered a collision with an uncharted pinnacle of rock. The lower bow casing and outboard end of the torpedo tube was damaged, jamming a warhead of a torpedo in the tube. On arrival in the dockyard, a survey was undertaken and she was moved into the floating dock in a quiet corner of the yard, where the torpedo was removed in a difficult and dangerous operation for which, it was said, there was no lack of volunteers.[5]

HMS *Hermes* was hurriedly prepared at Portsmouth to sail as the flagship of the Falklands task force on 5 April 1982, and here makes a triumphant return from the conflict to an enthusiastic reception on 21 July 1982.

The Run-Down Plan

With the successful end of the Falklands emergency, the ships returned but the rundown of the yard began. In 1984, the Royal Dockyard was renamed the Naval Base, and the Fleet Maintenance and Repair Organisation (FMRO) took over a reduced volume of ship refitting and repair. A swath of ships that were to have been withdrawn under the Defence Review were reprieved, including *Invincible*, *Fearless*, *Intrepid*, *Bristol*, *Glamorgan*, *Fife* and three Leander class frigates. The capacity needed for the repair and maintenance of the fleet was reappraised and 1,500 jobs were added to the projected Portsmouth establishment. The new total

of 2,800 would include 2,050 industrial and 750 non-industrial employees, to be achieved by 1 October 1984. The increase to 2,800 was contingent on the acceptance of flexible working practices, which was obtained from the trades unions. Management wanted flexible working to overcome the persistent problem of demarcation disputes which had dogged the dockyard. The new FMRO was under the control of the Commander-in-Chief, Fleet, represented by a Naval Base Commander (of rear admiral rank), but was under the professional guidance of the Chief Executive, Dockyards, insofar as Docking and Essential Defects periods and refits were concerned.[6]

In March 1983, a buildings closure plan was issued. Boathouses 5, 6 and 7 and the Block Mills would be transferred for heritage use, whilst the Iron and Brass Foundry, Nos 1 and 2 Smitheries, Welding Shop, Light Plate Shop, No. 8 Combined Workshop, MED Factory, Tool Room, No. 1 Electrical Shop and the Pipe Laundry would close. The Apprentice Training Centre would become a Hampshire County Council Training Centre. The number of machine tools would be reduced from 3,050 to about 1,000. The number of dockside cranes would reduce from sixty to thirty-six, and the 250-ton crane would be dismantled (this started on 8 December 1983 and took six months). The number of floating cranes would reduce from three to one. No. 2 Basin would have the caisson removed to make it a tidal basin, for use by minehunters transferred from HMS *Vernon*. Seven dry docks, Nos 1, 3, 4, 5, 6, 7 and 10, would be closed and the floating dock AFD 26 would move to Rosyth (it was towed away on 9 May 1984). Submarine dry-docking would be transferred from AFD 26 to No. 9 Dock. All this work was completed by the end of 1984, except for the closure of No. 10 Dock, which

did not occur until March 1989. The operational part of FMRO was now concentrated at the North Corner and around Nos 2 and 3 Basins. Seven dry docks and the four locks were retained, whilst the number of major workshops had been reduced from nineteen to nine.[7] The MED Factory was converted into Store No. 100 in 1986, extended in 1994, and in 1999 was equipped with automated materials handling systems.[8]

On 7 April 1983, the issue of redundancy notices recommenced and the process ended on 1 October 1984. In the industrial occupations (not including apprentices), there were 414 compulsory and 774 voluntary redundancies, 288 retirements (or deceased), 148 discharged at own request, 419 transfers (including 106 out of the area), 175 'job release' and 169 'services no longer required or medical discharge'. A total of 619 apprentices were discharged at the end of their apprenticeships, whilst thirty-four were retained as craftsmen, six transferred to other yards, thirty-nine technician apprentices were retained and twenty-seven apprentices from other establishments were returned to their own employer. In the non-industrial occupations, there were 419 retirements (188 of which were early), forty-nine medical discharges, sixteen deceased, ninety-six resignations, 493 transfers (including 279 out of area), four 'job release' and one dismissal. During the rundown, 139 non-industrial employees were recruited or promoted from industrials (including the thirty-nine technician apprentices), so that by 1 October 1984 there were a total of 2,947 employees in the yard.[9]

Recollections

There was a feeling of anger and helplessness amongst the workforce when the cuts were announced. For some time the view that dockyard work meant a job for life, which had prevailed up to the 1960s, had no longer been valid. Some of the anger was vented on John Nott, and also, during the minister's visit, the Pot Admiral was thrown to the ground.[10] One retired dockyardman recalled:

When the news reached everybody ... it was a bombshell! Men who had worked in the place all their lives, from a boy to a man, some with almost 40 to 45 years of proud service and a gift of craftsmanship that was second to none, were to be thrown on the

The 4½in gun of the destroyer *Bristol* is lifted whilst she is in refit in No. 15 Dock, February 1985. (Portsmouth Royal Dockyard Historical Trust)

scrapheap. Others, who had just joined, would be the case of last in first out. All of us felt very bitter at being betrayed, for all the many, many years of sacrifice, toil, sweat, labour and above all, pride, in producing some of the finest craftsmanship ever to repair a warship in refit … Everybody was worried to death – who would get the chop from management downwards.

This led to a mass meeting of employees, union leaders and shop stewards, and a downing of tools at 10.30 a.m., when the workforce walked out.[11] For some their hopes of a dockyard career were dashed: grandmother Jessie Worrall remembered, 'In 1981 more than 1,000 boys sat the craft apprentices' exam. My second grandson was one of the 168 selected. Three weeks later he received a letter saying there were no apprenticeships because of the run-down.'[12]

When the Falklands crisis brought temporary relief from the threatened redundancies, the workforce responded positively. 'There was no resentment felt over the redundancy issue, you were glad to be part of something again and to feel useful,' said a former joiner. An engine fitter recalled, 'Whilst the Falklands was on even the old demarcations were set aside. Everyone just buckled down … no trouble with the unions.' Yet the secretary of the Dockyard Whitley Committee said, 'The whole affair is scandalous. The government is rubbing salt into our wounds. Mrs Thatcher has kicked us in the teeth and now expects us to bail her out of trouble.' And bail the government out they did: The News reported:

For nearly two months dockyard craftsmen, workers in stores and armament depots, drivers, clerks and canteen staff laboured day and night. Ten thousand men and women were absorbed in a prodigious team effort for Britain … Men worked from 7 a.m. to 2 a.m. the following day … they had to be told to take a tea break.[13]

A former slinger felt, 'Money always buys loyalty, patriotism call it what you like … there was a chance to earn some money before being chucked out.' But a retired shipwright recollected, 'There was a job to be done, the country was in trouble, we had to get them out. It was patriotism not self-interest.' A redundant engine fitter said there was some hope, adding, 'You felt that you showed the government just what we were capable of and just what they would be losing.'[14] But apparently the government did not relent. The News reported:

Portsmouth civil and union leaders are furious at Defence Secretary John Nott's refusal to give ground over his plans to run down the dockyard. Despite the evidence of the dockyard's magnificent efforts to get the Falklands task force to sea so quickly, Mr Nott remains intransigent.[15]

However, as the former general manager who became Constructor Captain FMRO remembered:

As a result of the Falklands government rethink, one outcome was that the trade unions could have 2,800 jobs provided they signed up to flexible working practices. We got our 2,800, we got 500 naval people and the deal was that people in green overalls would work next to people in blue overalls, next to contractors, people in white overalls. We had a new productivity scheme attached to flexible working practices and the Treasury liked that.[16]

All this was part of the transformation from royal dockyard to naval base with support facilities that would eventually be privatised. On 1 October 1984, The Times carried an anonymous entry in the obituary section. It read, 'HM Royal Dockyard Portsmouth passed peacefully away at 12 o'clock last night after nearly 800 years of faithful service. It will be sadly missed by many.'[17]

The old order changeth yielding place to the new ...

Alfred, Lord Tennyson (1809–1892), 'The Passing of Arthur', (1869)

Following the 1981 Defence Review, the dockyards at Chatham and Gibraltar were closed. Devonport and Rosyth continued as the main refitting and repair yards, whilst Portsmouth had its subsidiary role in support and repair work on operational warships. The size of the fleet continued to decline continuously over the ensuing decades. By 2018, it comprised just one aircraft carrier, two assault ships, nineteen destroyers and frigates and ten submarines, and of course this downsizing reduced the workload in the yard.

The first dry-docking of a nuclear submarine at Portsmouth took place on 9 October 1987 when *Swiftsure* docked in C Lock. A glimpse of the work of the yard in the mid 1980s is given by statistics showing the number of vessels of all types docked between 1 October 1984 (when FMRO was formed) and 16 March 1989, a period of just under four-and-a-half years: eight aircraft carriers, thirty-seven destroyers, sixteen frigates, sixty-three submarines, ten landing craft, sixteen minehunters, seven support ships, thirty-nine small auxiliaries and six caissons were docked, a total of 200. Thereafter the workload continued to decline, for between 17 March 1989 and December 1995, a period of over six-and-a-half years, there were 245 dry-dockings.[1]

On 28 February 1998, the work of the Fleet Maintenance and Repair Organisation, i.e. warship repair, maintenance and refitting, was contracted out to Fleet Support Ltd (FSL), a private sector joint venture between BAE Systems and the VT Group. The company's work expanded to include the refitting of commercial ships such as cross-Channel and Isle of Wight ferries, British Antarctic Survey ships and the modernisation of redundant Royal Navy ships which have been sold to other navies. There was the occasional warship refit: in April 1999, *The News* announced 'Warship Refits back in the City' after Fleet Support Ltd won a £1m contract to refit HMS *Endurance*.[2]

In October 1998, a £2.5m scheme was announced to straighten and improve the jetties on the west side of the yard, creating Victory Jetty in place of Pitch House Jetty, and combining the North Railway Jetty with Sheer Jetty.[3] The Victorian railway shelter was moved in March 2004 from the South Railway Jetty to the north-east corner of the jetty overlooking the north Camber as work to strengthen the South Railway Jetty was planned.[4] Nos 7 and 10 Docks were filled in to provide more car parking space in 1991–92. A new Naval Base Commander's HQ, Victory Building, was built in 1993. In 2010, the Old Iron Foundry was restored and converted into BAE Systems offices.[5]

In 2003, after a gap of thirty-five years, shipbuilding returned to the dockyard, following the signing in 2002 of a 125-year lease of part of the naval base by the VT Group. This group (formerly Vosper Thornycroft) relocated its shipbuilding activity from Woolston, near Southampton, to the base, and opened a new facility which contained two new large enclosed building sheds on the site of the former No. 13 Dock. The largest of these was the Ship Assembly Hall, of 130m length, in which complete ships or large modules of ships were constructed from sub-units fabricated in the Unit Construction Hall, which was situated alongside the Ship Assembly Hall. Nearby was the Steelwork Production Hall, which was converted from an existing dockyard building to supply the required steelwork sections and panels to the two construction halls. The facility would build modules for the Navy's Type 45 destroyers and Queen Elizabeth class aircraft carriers.

The first vessel to be built at the new Portsmouth facility was the barge *Woolston*, of 1,350 tons, which was used to transport the VT-built bow sections, masts and funnels of the Type 45 destroyers to the BAE shipyard on the Clyde. She was named and rolled out of the main ship assembly hall on 21 June 2004. In June 2006, the first complete new ship to be built at Portsmouth for nearly forty

An Oberon class submarine refitting in No. 9 Dock. (Portsmouth Royal Dockyard Historical Trust)

The Portsmouth-built patrol vessel *Clyde* leaving Portsmouth Harbour during her period of trials and work-up in 2007. On completion of these, she was deployed for continuous service in the Falkland Islands.

years, the Royal Navy's offshore patrol vessel *Clyde*, was rolled out of the assembly hall for final fitting out in No. 14 Dock. She was commissioned on 30 January 2007 for service in the Falkland Islands.

In 2007, the number of civilian MOD employees in HM Naval Base and HMS *Nelson* (the naval barracks) was 1,786. Within the dockyard confines there were also the following numbers of private sector employees: defence prime contractors (VT Ltd, Fleet Support Ltd and BAE Systems), 2,478; sub-contractors, 1,111; Heritage, 245.[6]

On 1 July 2008, VT created a joint venture with BAE Systems known as BVT Surface Fleet, which also subsumed Fleet Support Ltd. On 30 October 2009, BAE acquired the VT Group's share of the joint venture in a rationalisation of British warship-building capacity, and the operation was renamed BAE Systems Surface Ships Ltd. On 1 January 2011, this combined with the company's submarine-building arm to become BAE Systems Maritime. Between 2008 and 2013, the Portsmouth facility built an offshore patrol vessel for Trinidad and Tobago (this ship, *Port of Spain*, was sold to Brazil after completion and renamed *Amazonas*) and three corvettes for the Royal Navy of Oman, as well as sections for the two new large aircraft carriers (*Queen Elizabeth* and *Prince of Wales*).

In 2013, it was announced that BAE's shipbuilding would be concentrated at Glasgow, said to be due to a reduction in activity levels following the Type 45 and aircraft carrier projects, but widely believed to be a politically inspired decision to strengthen the case for Scotland remaining in the United Kingdom to influence the impending referendum vote. Consequentially, after its short renaissance, warship-building at Portsmouth ended again with the loss of 940 direct jobs and 170 agency staff. On 18 August 2014, the last section built by BAE at Portsmouth, a 6,000-ton module for *Prince of Wales*, left Portsmouth on a barge for Rosyth. The shipbuilding site was handed back to the Ministry of Defence, who were seeking a new tenant, but in May 2016 it was reopened as a BAE Systems specialist refit centre for small ships, the Vernon Complex, allowing 'deep maintenance' to be carried out under cover, speeding up refit times. Work there started with the refit and re-engining of the minehunter *Brocklesby*.

Portsmouth Naval Base covers 300 acres of land plus 62 acres of basin, with 3 miles of waterfront.[7] It is one of three Royal Navy bases – the other two being Devonport and Clyde (principally Faslane). The operational work associated with the maintenance,

The Type 45 destroyer *Diamond* during an assisted maintenance period alongside the North-West Wall in April 2015.

repair and refitting of the Navy's ships is centred around No. 3 Basin with the associated dry docks (Nos 9, 11, 12, 14 and 15) and A–D Locks, and is now managed by BAE Systems. When Victory Jetty was extended, the former entrance to No. 6 Dock was blocked.[8] In 2014, the base had a labour force of 3,200. It is the home to over half of the surface fleet, including the two new large aircraft carriers, all of the Type 45 destroyers, half of the Type 23 frigates, half of the mine countermeasure vessels, the offshore patrol vessels and a number of small training vessels.

To accommodate the new carriers, a £100 million upgrade of the base was started in 2013. This included dredging a deeper channel on both sides of the harbour mouth, providing a deep water mooring for the carriers off Stokes Bay, rebuilding the Middle Slip Jetty, improving Fountain Lake Jetty, the South Railway Jetty and the Victory Jetty, installing new cranes and power supply, improving access to No. 3 Basin with new caissons and upgrading the base's facilities. The eighteen-month contract, costing £34m, to turn the 90-year-old structure at Middle Slip Jetty into modern berthing facilities for the carriers was awarded to Volker-Stevin, and work began in July 2015. In June 2015, the MOD awarded a contract worth £31 million to Fareham-based Boskalis Westminster Ltd for work to ready the harbour for the ships. To prepare for the arrival of

the main dredging vessel, the crane barge *Strekker* began removing debris from the Harbour, including nineteenth-century anchors, and dredging started in December 2015, uncovering amongst other things wartime bombs. On completion of all the work, the jetty was renamed the Princess Royal Jetty by Princess Anne at a ceremony there on 20 March 2017. The more modern Victory Jetty has also been improved, so that both 65,000-ton carriers can be in port simultaneously. HMS *Queen Elizabeth* arrived in Portsmouth for the first time on 17 August 2017 and berthed at the Princess Royal Jetty. The upgrade appears to assure the future of the base, which is still the city's largest direct and indirect employer. Within the base is the Royal Naval Supply Depot, which holds and distributes 80 per cent of the non-explosive stores used by the surface fleet. In December 2015, a £13.5 million contract with BAE Systems for sixty new Pacific 24 rigid hulled inflatable boats was announced. Designed by BAE Systems and built in Portsmouth Dockyard, they form a vital part of the Royal Navy, acting as ship-to-ship and ship-to-shore transfer, and rapid response craft in fast rescue, anti-piracy and counter-narcotics missions.

The International Festival of the Sea

By the 1990s, Navy Days was no longer an annual event: instead, Navy Days at Portsmouth were held every other year, alternating with Plymouth. In 1998, 2001 and 2005, an International Festival of the Sea was held, hosting not only British warships but also foreign warships and tall ships. Huge crowds were attracted, as many as 250,000 according to the organisers. If so, this exceeded even the popular pre-war Navy Weeks, and was ten times the attendance of the most recent Navy Days at Portsmouth, convincing the naval authorities of the value of the event. Britain's first International Festival of the Sea was held at Bristol in 1996, with the *Matthew* replica as a centrepiece – her construction was inspired by the 500th anniversary of John Cabot's voyage to Newfoundland in 1497. The next festival was in 1998 at Portsmouth Dockyard. The tall ships present included the Russian *Sedov*, *Kruzenshtern* and *Mir*, the Polish *Iskra*, the Argentinian *Libertad* and the replicas *Grand Turk* and *Matthew*. Amongst the warships, which were berthed separately from the tall ships, were *Invincible*, *Fearless*, *Birmingham*,

Liverpool, Manchester, Nottingham, Iron Duke and *Marlborough*. Access to most areas of the dockyard was possible – giving a rare opportunity to see both historic buildings and modern facilities.

The success of the event led to an even larger International Festival in 2001, when nearly 1,000 ships and small boats attended, manned by around 15,000 sailors. It was the largest public event of the year in Britain, and the largest maritime event in Europe. Tall ships and warships were intermixed along the jetties. The line-up of tall ships was impressive, including the Norwegian *Christian Radich, Sorlandet* and *Statsraad Lehmkuhl*, the Russian *Sedov, Mir* and *Shtandart*, the Polish *Dar Mlodziezy, Iskra* and *Pogoria*, the Italian *Palinuro*, the Brazilian *Cisne Branco*, the Mexican *Cuauhtemoc* and the British *Prince William, Royalist* and *Matthew*. Amongst the warships were *Illustrious, Nottingham, Exeter, York, Southampton, Grafton, Richmond, Westminster* and *Endurance*, plus eight foreign ships including USS *Winston Churchill* and the French *De Grasse*.

After being held at Edinburgh in 2003, the event returned to Portsmouth in 2005 as part of the bicentennial Trafalgar anniversary. The festival was preceded by an International Fleet Review at Spithead, attended by ships of thirty-five navies as well as a host of tall ships, merchant and private vessels; at the end of the review, many of these proceeded to Portsmouth Harbour to join the International Festival of the Sea. They included HM ships *Illustrious, Nottingham, Cumberland, St Albans, Marlborough, Richmond* and *Endurance*, RFA *Fort Victoria*, warships and auxiliaries from twenty-one navies and thirty tall ships, amongst which were *Eagle, Amerigo Vespucci, Gloria, Cisne Branco, Tarangini, Dewaruci, Sagres, Shabab Oman, Sorlandet, Mir, Jadran, Lord Nelson, Prince William, Capitan Miranda, Dar Mlodziezy, Grand Turk, Tenacious* and *Artemis*.

In 2008, Navy Days was restyled as Meet Your Navy. Areas of the dockyard that were normally closed to the public were again open, enabling the public to meet the crews of modern naval ships and explore the dockyard. The dockyard was split into past, present and future zones, with a packed programme including a field gun display, helicopter search and rescue and fly pass displays, a performance by the Band of the Japanese Navy, Georgian, Victorian and First World War naval and military re-enactments, the Royal Navy Historic Flight fly pass, a live music festival and the Royal Marines Band Beating the Retreat. The ships open to the public included naval ships from Brazil, Russia, Japan, Denmark and France, as well as British warships including *Illustrious, Gloucester, Campbeltown, Argyll, Richmond, Enterprise* and RFA *Largs Bay*. Another Meet Your Navy was held in 2010.

Portsmouth Historic Dockyard

In 1985, the Portsmouth Naval Base Property Trust (PNBPT) was established by the Ministry of Defence and Portsmouth City Council to take on the lease of the heritage area of the dockyard that was to become Portsmouth Historic Dockyard. It took responsibility for the long-term preservation of the historic south-west corner of Portsmouth Naval Base which had been released under the 1981 Defence Review, operating as a property development agency specialising in the reuse of historic buildings.

The buildings and docks within the heritage area include Nos 9–12 Storehouses, Nos 4–7 Boathouses, Nos 1–5 Docks, the Block Mills, Visitor Reception Area, Porters' Lodge, Porters' Garden, Cell Block, Victory Gate and Dockyard Wall, No. 19 College Road, Old Pay Office, Landing Stage, Mast Pond and Victory Gallery.

Tall ships alongside at the 2005 International Festival of the Sea.

The main entrance to the Historic Dockyard, Victory Gate dates back to 1711 and features a plaque on the right-hand wall which marks the visit of Queen Anne in the same year. Formerly Main Gate, it was widened during the Second World War to allow large vehicles to enter, and the wrought iron scrolled arch and lantern that spanned the two pillars was lost. The Porter's Garden was restored between 2000 and 2003, using the principles of eighteenth-century garden design, and reopened in April 2003.[9] To see the garden, you should turn right immediately beyond the Porter's Lodge after entering the Victory Gate. The garden contains the statue of William III, and at its entrance is a statue of Robert Falcon Scott. William III's statue, depicting him as a Roman emperor, was presented to the principal officers of the yard in 1718 by Richard Norton, of Southwick, who supplied timber to the yard, and was erected in the parade ground facing Long Row.[10] The bronze statue of Captain Scott, the Antarctic explorer, was subscribed by naval officers of the Portsmouth command and officers of the dockyard and unveiled by the wife of the C-in-C, Portsmouth, on 26 February 1915.[11]

In May 1992, the restoration of the clock tower on No. 10 Storehouse was completed, replacing the one destroyed by bombing in 1941. Together with Nos 11 and 12 Storehouses and the Victory Gallery, it is now home to the National Museum of the Royal Navy, Portsmouth (formerly the Royal Naval Museum). Part of No. 7 Boathouse houses the Dockyard Apprentice exhibition, opened in 1994, which tells the story of dockyard life in 1911, when the Dreadnoughts were being constructed. The exhibition invites the visitor to become an apprentice for the day, clocking in at the Victory Gate and learning all about the crafts and skills which once made Portsmouth Dockyard the greatest naval dockyard in the world. No. 6 Boathouse has housed an interactive display, Action Stations, but in 2016 it was announced that it would be home to the Royal Marines Museum, relocated from Eastney. This traces the history of the Marines back to 1664 and includes all of the ten Victoria Crosses won by Royal Marines.

No. 4 Boathouse has been restored and was opened in 2015 as the Boatbuilding & Heritage Skills Training Centre, delivering practical courses in traditional boatbuilding and related skills. Open to the public, it also houses a hands-on exhibition of historic naval boats, and a mezzanine-level brasserie with views across Portsmouth Harbour. It is a triumph of conservation and reuse

The statue of Robert Scott in the Historic Dockyard. It was sculpted by Kathleen Lady Scott in 1915 and commemorates her husband's journey to the South Pole and death in 1912.

Boatbuilding and repair in No. 4 Boathouse, now a training centre for heritage boatbuilding skills.

assault landing craft *F8*, which was built by Camper & Nicholsons, Gosport, in 1959–60; and a number of other small historic boats. Additionally, two large Second World War military powerboats are kept afloat and are sea-going: *MGB81*, a motor gunboat built by BPB in 1942; and *HSL102*, an air sea rescue launch built in 1937, also by BPB. The National Museum of the Royal Navy owns and operates the 50ft steam pinnace *199*, built at Cowes by J. Samuel White in 1911, probably for the battleship HMS *Monarch*.

The historic ships preserved in the Historic Dockyard are major tourist attractions. In No. 2 Dock is HMS *Victory* (1765), where she has lain since January 1922. In March 2012, responsibility for the ship passed from the Ministry of Defence to the National Museum of the Royal Navy, and £50 million was allocated for a major conservation programme which will take fifteen years to complete. In No. 1 Dock is HM monitor *M33* (1915), which has been there since 23 April 1997. Conservation work culminated in her being opened to the public in August 2015. The dry dock is no longer connected to the harbour. In No. 3 Dock lie the remains of *Mary Rose* (1511), which were raised from the seabed on 11 October 1982. The new Mary Rose Museum was opened here on 31 May 2013. The building takes the form of a finely crafted 'jewellery box', with the hull at its centre and galleries running the length of the ship, each corresponding to a deck level on the ship. Many of the objects recovered from the site, forming the most comprehensive collection of Tudor artefacts in the world, are displayed in the galleries by theme to help reveal some of the personal stories of life on board. In June 1987, the newly restored steam and sail frigate HMS *Warrior* (1860) returned to Portsmouth, her old base, where Portsmouth City Council had spent £1.5m on a new jetty and supporting facilities for her. *Warrior's* position on a jetty near the Hard allows her to be seen to good effect and attracts the attention of visitors to the Historic Dockyard and passengers on the Gosport ferry.

of a redundant dockyard building, especially as its original role – building and repairing small boats – has been revived. The historic boat collection includes the recently restored *Cyclops*, a 42ft 1916 rowing launch (later a workboat) belonging to the Portsmouth-built First World War battleship *Royal Sovereign*; a 31ft steam cutter built in Cowes by J. Samuel White in 1896 and carried by the royal yacht *Osborne*; the 1956 dockyard harbour launch *D49*; the 45ft picket boat *Green Parrot*, also built in 1956, which was latterly used as the C-in-C Portsmouth's barge; the 41ft seaplane tender *ST1502*, built in 1942 by British Power Boat, Hythe (BPB); the 43ft

Name	Launched	Type	Tonnage[1]	Name	Launched	Type	Tonnage
Sweepstake	. . 1497	Ship	80	Royal James	29.6.1675	1st Rate	1,486
Mary Fortune	. . 1497	Ship	80	Vanguard	. . 1678	2nd Rate	1,357
Mary Rose	. . 1509	Ship	500	Eagle	. . 1679	3rd Rate	1,053
Peter Pomegranate	. . 1510	Ship	450	Expedition	. . 1679	3rd Rate	1,116
Jennet	. . 1539	Ship (galleasse)	180	Ossory	. . 1682	2nd Rate	1,307
Portsmouth	. . 1649	4th Rate	422	Coronation	. . 1685	2nd Rate	1,346
Laurel	. . 1651	4th Rate	489	Nonsuch	. . 1686	Hoy	95
Martin	. . 1652	6th Rate	127	Portsmouth	1.5.1690	5th Rate	412
Sussex	. . 1652	6th Rate	600	Norwich	16.7.1691	5th rate	616
Bristol	. . 1653	4th Rate	532	Russell	3.6.1692	2nd Rate	1,177
Marigold	. . 1653	Hoy	42	Forester	2.11.1693	Hoy	125
Lyme	. . 1654	4th Rate	764	Weymouth	8.8.1693	4th Rate	673
Dartmouth	. . 1655	6th Rate	260	Fly	.3.1694	Advice boat	73
Wakefield	. . 1655	6th Rate	313	Mercury	19.3.1694	Advice boat	78
Chestnut	. . 1656	Ketch	81	Newport	7.4.1694	6th Rate	253
Monck	7.1.1661	3rd Rate	684	Scout	13.7.1694	Advice boat	38
Royal Oak	26.12.1664	2nd Rate	1,021	Lichfield	4.2.1695	4th Rate	682
Portsmouth	. . 1665	Ketch	90	Shrewsbury	6.2.1695	2nd Rate	1,257
Portsmouth	. . 1667	Sloop	43	Express	. . 1695	Advice boat	77
Nonsuch	.12.1668	4th Rate	359	Postboy	14.4.1695	Advice boat	77
St Michael	. . 1669	2nd Rate	1,080	Fly	11.2.1696	Ketch	70
Saudadoes	. . 1670	Sloop	83	Association	1.1.1697	2nd Rate	1,459
Cleveland	. . 1671	Royal yacht	107	Exeter	26.5.1697	4th Rate	949
Phoenix	.3.1671	5th Rate	368	Looe	15.10.1697	5th Rate	390
Royal James	. . 1671	1st Rate	1,426	Seaford	15.10.1697	6th Rate	248
Greyhound	. . 1672	6th Rate	180	Nassau	2.8.1699	3rd Rate	1,080
Prevention	. . 1672	Sloop	46	Swift	31.10.1699	Sloop	65
Cutter	. . 1673	Ketch	46	Woolf	31.10.1699	Sloop	65
Hunter	. . 1673	Sloop	46	Portsmouth	11.1.1703	Royal yacht	50
Invention	. . 1673	Sloop	28	Squirrel	14.6.1703	6th Rate	258
Isle of Wight	. . 1673	Yacht	31	Roebuck	5.4.1704	5th Rate	494
Navy	. . 1673	Yacht	74	Squirrel	28.10.1704	6th Rate	260
Royal Charles	10.3.1673	1st Rate	1,443	Nassau	9.1.1706	3rd Rate	1,104

Name	Launched	Type	Tonnage	Name	Launched	Type	Tonnage
Hastings	2.10.1707	5th Rate	533	Sphinx	25.10.1775	6th Rate	431
Truelove	22.11.1707	Hoy	76	Cygnet	24.1.1776	Sloop	301
Sapphire	3.9.1708	4th Rate	686	Swift	9.1.1777	Sloop	303
Bolton	19.7.1709	Yacht	42	Lion	3.9.1777	3rd Rate	1,378
Fowey	7.12.1709	5th Rate	528	Eurydice	26.3.1781	6th Rate	521
Seahorse	13.2.1711	6th Rate	282	Crocodile	24.4.1781	6th Rate	519
Solebay	21.4.1711	6th Rate	272	Warrior	18.10.1781	3rd Rate	1,642
Launceston	10.10.1711	5th Rate	528	St George	4.10.1785	2nd Rate	1,950
Success	30.4.1712	6th Rate	275	Serpent	3.12.1789	Sloop	321
Spy	9.12.1721	Sloop	103	Fury	2.3.1790	Sloop	323
Hayling	. . 1729	Hoy	126	Swift	5.10.1793	Sloop	329
Centurion	6.1.1732	4th Rate	1,005	Prince of Wales	28.6.1794	2nd Rate	2,024
Shark	7.9.1732	Sloop	201	Dreadnought	13.6.1801	2nd Rate	2,123
Worcester	20.12.1732	4th Rate	1,061	Grampus	20.3.1802	4th Rate	1,115
Victory	23.2.1737	1st Rate	1,921	Alexandria	18.2.1806	5th Rate	663
Portsmouth	. . 1742	Royal yacht	83	Bulwark	23.4.1807	3rd Rate	1,940
Tilbury	20.7.1745	4th Rate	1,124	Brazen	26.5.1808	6th Rate	422
Wasp	4.7.1749	Sloop	140	Podargus	26.5.1808	Brig	254
Hazard	3.10.1749	Sloop	140	Zephyr	28.4.1809	Brig	253
Grafton	29.3.1750	3rd Rate	1,414	Primrose	22.1.1810	Brig	383
Newcastle	4.12.1750	4th Rate	1,053	Pyramus	22.1.1810	5th Rate	920
Fly	9.4.1752	Sloop	140	Boyne	3.7.1810	2nd Rate	2,155
Chichester	4.6.1753	3rd Rate	1,401	Hermes	22.7.1811	6th Rate	511
Gibraltar	9.5.1754	6th Rate	430	Childers	8.7.1812	Brig	384
Neptune	8.12.1756	2nd Rate	1,799	Lacedaemonian	21.12.1812	5th Rate	1,073
Neptune	17.7.1757	2nd Rate	1,798	Grasshopper	17.5.1813	Brig	385
Dorsetshire	13.12.1757	3rd Rate	1,436	Vindictive	23.11.1813	3rd Rate	1,758
Chatham	25.4.1758	4th Rate	1,053	Icarus	18.8.1814	Brig	234
Britannia	19.10.1762	1st Rate	2,091	Pallas	13.4.1816	5th Rate	951
Asia	16.5.1764	3rd Rate	1,364	Pitt	13.4.1816	3rd Rate	1,751
Warwick	28.2.1767	4th Rate	1,073	Waterloo	16.10.1818	3rd Rate	2,056
Ajax	23.12.1767	3rd Rate	1,615	Delight	10.5.1819	Brig	237
Elizabeth	17.10.1769	3rd Rate	1,617	Cygnet	11.5.1819	Brig	237
Worcester	17.10.1769	3rd Rate	1,380	Britomart	24.8.1820	Brig	237
Falcon	15.6.1771	Sloop	302	Prince Regent	30.5.1820	Yacht	282
Medina[2]	.8.1772	Yacht	66	Minerva	13.6.1820	5th Rate	1,082
Princess Royal	18.10.1773	2nd Rate	1,973	Jasper	26.7.1820	Brig	237
Berwick	18.4.1775	3rd Rate	1,623	Ranger	7.12.1820	6th Rate	502

Name	Launched	Type	Tonnage	Name	Launched	Type	Tonnage
Martin	18.5.1821	Sloop	461	Electra	17.2.1837	Sloop	462
Rose	1.6.1821	Sloop	398	Hazard	21.4.1837	Sloop	431
Plover	30.6.1821	Brig	237	Termagant	26.3.1838	Brigantine	231
Ferret	12.10.1821	Brig	237	Indus	16.3.1839	2nd Rate	2,098
Arrow	14.3.1823	Cutter	157	Queen	15.5.1839	1st Rate	3,104
Philomel	28.4.1823	Brig	237	Stromboli	27.8.1839	Paddle sloop	967
Royalist	12.5.1823	Brig	231	Bittern	18.4.1840	Brig	484
Tweed	14.4.1823	6th Rate	500	Rapid	3.6.1840	Brig	319
Carnatic	21.10.1823	3rd Rate	1,790	Driver	24.12.1840	Paddle sloop	1,058
Champion	31.5.1824	Sloop	456	Thunderbolt	13.1.1842	Paddle sloop	1,058
Orestes	31.5.1824	Sloop	460	Albatross	28.3.1842	Brig	484
Leveret	19.2.1825	Brig	232	Frolic	23.8.1842	Sloop	511
Volage	19.2.1825	5th Rate	516	Firebrand	6.9.1842	Paddle frigate	1,190
Musquito	19.2.1825	Brig	231	Eurydice	16.5.1843	6th Rate	908
Myrtle	14.9.1825	Brig	231	Sealark	27.7.1843	Brig	319
Princess Charlotte	11.11.1825	2nd Rate	2,443	Daring	2.4.1844	Brig	426
Challenger	14.11.1826	6th Rate	603	Osprey	2.4.1844	Brig	425
Columbine	1.12.1826	Sloop	492	Scourge	8.11.1844	Paddle sloop	1,128
Wolf	1.12.1826	Sloop	454	Centaur	6.10.1845	Paddle frigate	1,296
Sylvia	24.3.1827	Cutter	70	Rifleman	10.8.1846	Screw gun vessel	486
Sapphire	31.12.1827	6th Rate	604	Dauntless	5.1.1847	Screw frigate	1,453
President	20.4.1829	4th Rate	1,537	Leander[3]	8.3.1848	4th Rate	1,987
Favourite	21.4.1829	Sloop	434	Plumper	5.4.1848	Screw sloop	490
Fox	17.8.1829	5th Rate	1,080	Arrogant	5.4.1848	Screw frigate	1,872
Rapid	17.8.1829	Brig	235	Argus	15.12.1849	Paddle sloop	981
Recruit	17.8.1829	Brig	235	Furious	26.8.1850	Paddle frigate	1,287
Seaflower	20.5.1830	Cutter	116	Princess Royal[4]	23.6.1853	Screw 2nd Rate	3,129
Actaeon	. 1.1831	6th Rate	620	Marlborough	31.7.1855	Screw 1st Rate	4,000
Charybdis	27.2.1831	Brig	232	Shannon	24.11.1855	Screw frigate	2,667
Fanny	28.2.1831	Admiralty yacht	136	Royal Sovereign	25.4.1857	Screw 1st Rate	3,756
Neptune	27.9.1832	1st Rate	2,694	Bacchante	30.7.1859	Screw frigate	2,667
Racer	18.7.1833	Brig	431	Victoria	12.11.1859	Screw 1st Rate	4,127
Lynx	2.8.1833	Brigantine	231	Duncan	13.12.1859	Screw 1st Rate	3,727
Buzzard	23.3.1834	Brigantine	231	Prince of Wales	25.1.1860	Screw 1st Rate	3,994
Drake	25.3.1834	Dockyard lighter	109	Frederick William	24.3.1860	Screw 1st Rate	3,241
Hermes	26.6.1835	Paddle sloop	716	Rinaldo	26.3.1860	Screw sloop	951
Inconstant	10.6.1836	5th Rate	1,422	Chanticleer	9.2.1861	Screw sloop	950
Volcano	29.6.1836	Paddle sloop	720	Glasgow	28.3.1861	Screw Frigate	3,027

Name	Launched	Type	Tonnage	Name	Launched	Type	Tonnage
Cherub[5]	29.3.1864	Gunboat	330	Pallas	30.6.1890	3rd class cruiser	2,575
Royal Alfred	15.10.1864	Ironclad	4,068	Royal Arthur	26.2.1891	1st class cruiser	7,700
Helicon	31.1.1865	Dispatch vessel	837	Royal Sovereign	26.2.1891	Battleship	15,585
Minstrel[6]	16.2.1865	Gunboat	330	Crescent	30.3.1892	1st class cruiser	7,700
Netley[7]	22.7.1866	Gunboat	268	Centurion	3.8.1892	Battleship	10,500
Orwell[8]	27.12.1866	Gunboat	330	Fox	15.6.1893	2nd class cruiser	4,360
Bruizer[9]	23.4.1867	Gunboat	330	Eclipse	19.7.1894	2nd class cruiser	5,600
Danae	21.5.1867	Corvette	1,287	Majestic	31.1.1895	Battleship	14,900
Cromer[10]	20.8.1867	Gunboat	330	Prince George	22.8.1895	Battleship	14,900
Ringdove	4.9.1867	Gun vessel	467	Caesar	2.9.1896	Battleship	14,900
Avon	2.10.1867	Gun vessel	467	Gladiator	18.12.1896	2nd class cruiser	5,750
Cracker	27.11.1867	Gun vessel	467	Canopus	12.10.1897	Battleship	14,320
Elk	10.1.1868	Gun vessel	467	Formidable	17.11.1898	Battleship	15,000
Magpie	12.2.1868	Gun vessel	665	London	21.9.1899	Battleship	15,640
Sirius	24.4.1868	Corvette	1,268	Pandora	17.1.1900	3rd class cruiser	2,200
Swallow	16.11.1868	Gun vessel	664	Kent	6.3.1901	1st class cruiser	9,800
Dido[11]	23.10.1869	Corvette	1,277	Suffolk	15.1.1903	1st class cruiser	9,800
Plucky[12]	13.7.1870	Gunboat	196	New Zealand	4.2.1904	Battleship	16,350
Blazer	7.12.1870	Gunboat	254	Britannia	10.12.1904	Battleship	16,350
Comet	8.12.1870	Gunboat	254	Dreadnought	10.2.1906	Battleship	17,900
Devastation	12.7.1871	Turret ship	4,406	Bellerophon	27.7.1907	Battleship	18,600
Shah	9.9.1873	Frigate	6,250	St Vincent	10.9.1908	Battleship	19,560
Boadicea	16.10.1875	Corvette	3,913	Neptune	30.9.1909	Battleship	19,900
Inflexible	27.4.1876	Battleship	11,880	Orion	20.8.1910	Battleship	22,500
Bacchante	19.10.1876	Corvette	4,103	King George V	9.10.1911	Battleship	23,000
Canada	26.8.1881	Corvette	2,380	Iron Duke	12.10.1912	Battleship	25,000
Cordelia	25.10.1881	Corvette	2,380	Queen Elizabeth	16.10.1913	Battleship	27,500
Colossus	21.5.1882	Battleship	9,150	Royal Sovereign	29.4.1915	Battleship	27,500
Imperieuse	18.12.1883	Armoured cruiser	8,500	J1	6.11.1915	Submarine	1,210
Calliope	24.7.1884	Corvette	2,770	J2	6.11.1915	Submarine	1,210
Camperdown	24.11.1885	Battleship	10,600	K2	14.10.1916	Submarine	1,780
Trafalgar	20.8.1887	Battleship	12,590	K1	14.11.1916	Submarine	1,780
Nymphe	1.5.1888	Sloop	1,140	K5	16.12.1916	Submarine	1,780
Melpomene	20.9.1888	2nd class cruiser	2,950	Effingham	8.6.1921	Cruiser	9,750
Beagle	28.2.1889	Sloop	1,170	Murex	. .1922	Tanker (commercial)	8,887
Barrosa	16.4.1889	3rd class cruiser	1,580	Suffolk	16.2.1926	Cruiser	9,800
Vulcan	13.6.1889	Torpedo boat carrier	6,600	London	14.9.1927	Cruiser	9,850
Barham	11.9.1889	3rd class cruiser	1,803	Dorsetshire	29.1.1929	Cruiser	9,975

Name	Launched	Type	Tonnage	Name	Launched	Type	Tonnage
Comet	30.9.1931	Destroyer	1,375	Sirius	18.4.1940	Cruiser	5,600
Crusader	30.9.1931	Destroyer	1,375	Tireless	19.3.1943	Submarine	1,090
Nightingale	30.9.1931	Mining tender	298	Token	19.3.1943	Submarine	1,090
C307	. .1932	Tug	192	Thor[13]	18.4.1944	Submarine	1,090
C308	. .1932	Tug	192	Tiara[14]	18.4.1944	Submarine	1,090
Duncan	7.7.1932	Destroyer	1,400	Leopard	23.5.1955	Frigate	2,300
Skylark	15.11.1932	Mining tender	302	Rhyl	23.3.1959	Frigate	2,380
Neptune	31.1.1933	Cruiser	7,270	Nubian	6.9.1960	Frigate	2,300
Amphion	27.7.1934	Cruiser	6,908	Sirius	22.9.1964	Frigate	2,450
Exmouth	7.2.1936	Destroyer	1,474	Andromeda	24.5.1967	Frigate	2,500
Aurora	20.8.1936	Cruiser	5,270				

Ships Rebuilt at Portsmouth Royal Dockyard 1666-1748

Name	Launched	Type	Tons	Original build & former name	Name	Launched	Type	Tons	Original build & former name
Constant Warwick	21.6.1666	4th Rate	341	1645	Dreadnought	11.3.1723	4th Rate	932	1706
Bonaventure	. .1683	4th Rate	564	1663	Portland	25.2.1723	4th Rate	772	1693
Dover	12.12.1695	4th Rate	604	1654	Lowestoffe	18.12.1723	6th Rate	378	1687
Greenwich	. . 1699	4th Rate	785	1666	Kinsale	13.4.1724	5th Rate	607	1700
St George	. .1701	2nd Rate	1,470	1668 Charles	Southsea Castle	10.7.1724	5th Rate	553	1708
Isle of Wight	. .1701	Yacht	38	1673	Adventure	4.6.1726	5th Rate	598	1704
Restoration	22.1.1702	3rd Rate	1,045	1678	Princess Amelia	4.10.1726	3rd Rate	1,353	1693 Humber
Edgar	. .1702	3rd Rate	1,199	1668	Salisbury	30.10.1726	4th Rate	756	1707
Royal Katherine	23.2.1703	2nd Rate	1,395	1664	Oxford	10.7.1727	4th Rate	767	1674
Elizabeth	. . 1704	3rd Rate	1,152	1679	Aldborough	21.10.1727	6th Rate	375	1706
Captain	6.7.1708	3rd Rate	1,075	1678	Flamborough	21.10.1727	6th Rate	377	1707
Prince Frederick	16.8.1714	3rd Rate	1,116	1679 Expedition	Ipswich	30.10.1730	3rd Rate	1,142	1694
Montagu	26.7.1716	4th Rate	920	1654	Dunkirk	3.9.1734	4th Rate	965	1651
Nonsuch	29.4.1717	4th Rate	687	1696	Victory	23.2.1737	1st Rate	1,921	1695
Monmouth	3.6.1718	3rd Rate	1,174	1667	St George	3.4.1740	2nd Rate	1,655	1668 Charles
Kingston	9.5.1719	4th Rate	918	1697	Bedford	9.3.1741	3rd Rate	1,230	1698
Royal William	3.9.1719	1st Rate	1,918	1692 Prince	Princess Mary	5.10.1742	4th Rate	1,068	1704 Mary
Captain	21.5.1722	3rd Rate	1,131	1678	Sunderland	4.4.1744	4th Rate	1,123	1724
Lancaster	1.9.1722	3rd Rate	1,366	1694	Ramillies	8.2.1748	2nd Rate	1,685	1706
Canterbury	15.9.1722	4th Rate	963	1693					

NOTES AND REFERENCES

Abbreviations

BL	British Library.
NMM	National Maritime Museum, Greenwich.
NMRN	National Museum of the Royal Navy Library, Portsmouth.
PCL	Portsmouth Central Library.
PP	Parliamentary Papers.
PRDHT	Portsmouth Royal Dockyard Historical Trust.
TNA	The National Archives, Kew.

Chapter 1: Beginnings

1 Portus Adurni was the name of a Roman fort, the last of nine forts listed in a Roman military handbook of the late fourth century, but it is not certain that this was the one at Portchester; see Tomalin, D., *Footholds on the Shore*, in Drummond & McInnes, 2001, 67.
2 Winton, 1989, 12.
3 Tomalin, op. cit., 75.
4 Bingeman, J., 2015, King John's Dock, Scuttlebutt, 50, 54.
5 PRDHT, 1212: A Royal Dockyard is Born, Dockyard Timeline, http://portsmouthdockyard.org.uk/timeline/details/1212-a-royal-dockyard-is-born, accessed 8.3.17; Oppenheim, 1896, xiii.
6 MacDougall, 1982, 19; Rodger, 2004a, 53.
7 Carr Laughton, L.G., 1912, *Maritime History, A History of the County of Hampshire*: Vol. 5, Victoria County History, University of London, 359–407, cited by Coats, A., 2013, 'Portsmouth's First Dockyard', *Dockyards*, Naval Dockyards Society, 18:2, 3–4.
8 Bingeman, op. cit.
9 Page, W. (ed.), 1908, *The Liberty of Portsmouth and Portsea Island*, Vol. 3, Victoria County History, 177–92, http://www.british-history.ac.uk/report.aspx?compid=41952, accessed 14.11.2013, cited by Coats, A., op. cit.
10 Colvin, H.M. (ed.), 1963, *A History of the King's Works, The Middle Ages II*, London: HMSO, 988, cited by Coats, A., op. cit.

11 Rodger, 2004a, 102.
12 Winton, 1989, 19.
13 Oppenheim, 1896, 138; Patterson, 2008, 11–12.
14 Oppenheim, 1896, 143–59 (from TNA E315, E316); Lipscomb, 1967, 42–43.
15 Patterson, 2008, 11–12; Oppenheim, 1896, 143–59 (from TNA E315, E316).
16 Oppenheim, 1896, viii.
17 Patterson, 2008, 11–12.
18 Ibid.
19 Oppenheim, 1896, xxvi–xxvii.
20 Childs, 2007, 24–25.
21 Ibid., 28.
22 MacDougall, 1982, 30–31.
23 Winton, 1989, 22.
24 Archibald, 1972, 112, citing Derrick, 1806.
25 Winton, 1989, 22–23.
26 Ibid., 28.
27 Kitson, H., 1947, 'The Early History of Portsmouth Dockyard, 1496–1800', *The Mariners Mirror*, 33:4, 259.

Chapter 2: The Seventeenth-Century Dockyard

1 Goss, 1984, 14.
2 Ibid.
3 Kitson, H., 1947, 'The Early History of Portsmouth Dockyard, 1496–1800', *The Mariners Mirror*, 257–59; Lipscomb, 1967, 104–08; Winton, 1989, 38–39.
4 Winton, 1989, 40.
5 Ibid., 25.
6 Ibid., 40.
7 Coad, 2013, 147; Kitson, op. cit., 260–62.
8 Winton, 1989, 44–45.
9 Patterson, 2008, 23.

10   Archibald, 1972, 123–27, citing Derrick, 1806.

11   Patterson, 2008, 26.

12   Johns, A.W., 1925, 'Sir Anthony Deane', *The Mariner's Mirror*, 11:2, 164–93.

13   Patterson, 2008, 27.

14   Riley, 1985, 3; Patterson, 2008, 32.

15   Coad, 2013, 89.

16   Ibid., 90.

17   Ibid., and Riley, 1985, 5.

18   Riley, 1985, 5.

19   BL, 1698, Edmund Dummer, *A Survey and description of the principal harbours, with their accommodations and conveniences for erecting, moaring, secureing and refitting the Navy Royall of England … also an account of the nature and usefulness of the late erected new docks at Portsmouth.*

20   Coad, 2013, 89–91.

21   BL, 1698, op. cit.; Fox, C., 2007, 'The Ingenious Mr Dummer: Rationalizing the Royal Navy in Late Seventeenth-Century England', *Electronic British Library Journal*, 1–58, accessed 30.4.17.

22   Fox, 2007, op. cit..

23   Fox, 2007, op. cit., 1.

24   Fox, 2007, p. cit., 1–58; MacDougall, P., 2004, 'Edmund Dummer', *Oxford Dictionary of National Biography*, Oxford University Press; https://en.wikipedia.org/wiki/Edmund_Dummer_%28naval_engineer%29, accessed 30.4.2017.

25   Frigate Shtandart, https://www.shtandart.ru/en/frigate/history/original/, accessed 16.4.17.

26   MacDougall, 2012, 23.

Chapter 3: The Georgian Dockyard

1   Patterson, 2008, 32.

2   Riley, 1985, 5.

3   Coad, 2005, 25–26.

4   NMM, PFR/2/A, Officers' Houses, Long Row (1717 and 1719).

5   Patterson, 2008, 33.

6   NMM, PFR/2/A, Officers' Houses, Long Row (1717 and 1719).

7   Coats, A. et al, 2015, 140–41.

8   Portsmouth Naval Base Property Trust, Porter's Lodge, http://www.pnbpropertytrust.org/propertypage.asp?propid=6&pp=1, accessed 10.10.15.

9   Coad, 1989, 80–81.

10   Patterson, 2008, 34.

11   Riley, 1985, 3–4.

12   Coad, 1981, 17–18.

13   Patterson, 2008, 42.

14   TNA, ADM 7/662 FF 23R–34R cited in Knight, 1987, li, 138.

15   Riley, 1985, 7.

16   Ibid., 10.

17   Knight, 1987, lii.

18   Cock, R., 2007, 'At War with the Worm: The Royal Navy's Fight against the Shipworm and Barnacle, 1708–1793', *Transactions of the Naval Dockyards Society*, 3, 9–30.

19   Knight, 1987, 165.

20   Colledge & Warlow, 2006, 272.

21   Cock, op. cit., 9–30; Knight, 1987, 159–65; Knight, R., 'The Introduction of Copper Sheathing into the Royal Navy', *The Mariners Mirror*, 59:3, 299–309.

22   Knight, 1987, lv–lviii.

23   NMM, PDR/J/2; Knight, 1987, 163.

24   Derived from data in Knight, 1987, 159–65.

25   NMM, POR/D/19, POR/A/26, POR/G/1 and POR/F/15.

26   NMM, POR/D/19.

27   Goss, 1984, 20–21.

28   Riley, 1985, 7; Lewis, 1854, 24.

29   Coats, A. et al, 2015, 144.

31   Slight, 1828, 144, 157.

32   Riley, 1985, 12.

33   Coad, 2013, 182.

34   Patterson, 2008, 39.

35   NMM, PFR/2/A, Fires, 1760, 1770 and 1776; Patterson, 2008, 39.

36   NMM, PFR/2/A, Fires, 1760, 1770 and 1776.

37   Ibid.; Patterson, 2008, 39.

38   TNA, ADM 7/593.

39   Coad, 1989, 204.

40   TNA, ADM 7/593.

41   Slight, 1828, 136.

42   Ibid.; Lewis, 1854, 10.

43   TNA, ADM 7/593.

44   Warner, 2004, 152–56, 171.

45   Patterson, 2008, 42–43.

46   NMM, PFR/2/A, Fires, 1760, 1770 and 1776.

47   Coad, 1989, 133.

48   NMM, POR/G/1, POR/F/16, cited in Knight, 1987, 102–04.

49   NMM, POR/G/1, cited in Knight, 1987, 102–04.

50   Knight, 1987, xlix.

51   Wilkin, F.S., 2006, 'The Contribution of Portsmouth Royal Dockyard to the success of the Royal Navy in the Napoleonic War 1793–1815', *Transactions of the Naval Dockyards Society*, 1, 53.

52   Slight, 1828, 137.

53   Coad, 2005, 133.

54   Coats, A. et al, 2015, 131.

55   Ibid., 131–34.

56   Clark, C., 2008, 'Adaptive Re-use and the Georgian storehouses of Portsmouth – Naval Storage to Museum', *Transactions of the Naval Dockyards Society*, 4, 31.

57   Coats, A. et al, 2015, 130.

58   NMM, ADM 7/593.

59   Riley, 1985, 10.

60   Coad, 1989, 53–54; Coats et al, 2015, 132; PRDHT, http://portsmouthdockyard.org.uk/timeline/details/1785-st.-anns-church, accessed 2.1.16.

61   Patterson, 2008, 47.

62   Coad, 1989, 55.

63   NMM, PFR/2/A, Commissioner's House 1783–1785.

64   Coad, 2013, 163.

65   Horne, 1966; Colledge & Warlow, 2006, 373; The figurehead of *Victoria and Albert* 1899–1954 is positioned in front of the buildings of HMS *Nelson* barracks.

66   Samuel Bentham, Wikipedia, https://en.wikipedia.org/wiki/Samuel_Bentham, accessed 1.1.16; Coad, 2013, 71–72; Coad, 2005, 23–24.

67   Coad, 2013, 94.

68   Coad, 2005, 27–28.

69   Ibid., 32. Coad, 2013, 121–22.

70   Knight, 1987, lii.

71   Morriss, R., 2006, 'The Office of the Inspector General of Naval Works and Technological Innovation in the Royal Dockyards', *Transactions of the Naval Dockyards Society*, 1, 26.

72   Coad, 2005, 27–28; Brown, 2016, 167.

73   Morriss, 1983, 54.

74   Coad, 2005, 39–47.

75   Coad, 1989, 234.

76   Coats, A., 2006, 'The Block-mills: New Labour Practices for New Machines?', *Transactions of the Naval Dockyards Society*, 1, 68.

77   Wilkin, op. cit., 55.

78   Morriss, R., 2006, 'The Office of the Inspector General of Naval Works and Technological Innovation in the Royal Dockyards', *Transactions of the Naval Dockyards Society*, 1, 25.

79   Riley, R., 2006, 'Marc Brunel's Pulley Block-making Machinery: Operation and assessment', *Transactions of the Naval Dockyards Society*, 1, 85–92.

80   Coad, 2005, 49–52; Coats, A., 2006, op. cit., 74.

81   Coad, 2005, 99, 108.

82   Wilkin, op. cit., 55.

83   Coad, 2005, 49, 75–79; Coats, 2006, op. cit., 74, 78.

84   Coad, 2005, 44–47.

85   Morriss, R., 2006, 'The Office of the Inspector General of Naval Works and Technological Innovation in the Royal Dockyards', *Transactions of the Naval Dockyards Society*, 1, 24.

86   Ibid., 21–22; Morriss, R., 'Benthamism in the Royal Dockyards and British Public culture 1750–1850', *Transactions of the Naval Dockyards Society*, 4, 21–29.

87   Wilkin, op. cit., 47–50.

88   Ibid., 57.

89   Winfield, 2005, 24.

90   Ibid., 23, 251.

91   Goss, 1984, 23.

92   Wilkin, op. cit., 54.

93   Ibid., 54–55.

94   Thomas, J., 'Portsmouth Yard and Town in the Age of Nelson (1758–1805) – A Relationship Examined', *Transactions of the Naval Dockyards Society*, 1, 93–107.

95   Coad, 2013, 101.

96 The Times, 16.9.1825; Slight, 1828, 145–50; Lewis, 1854, 19–20.
97 Coad, 2013, 389–390; British Listed Buildings, Former School of Naval Architecture, http://www.britishlistedbuildings. co.uk/en-476698-former-school-of-naval-architecture-buil#. VsRIFOYaaJc, accessed 2.1.16.
98 Coats, A. et al, 2015, 120.
99 Evans, 2004, 20–25.
100 Winton, 1989, 103.

Chapter 4: Dockyard Administration

1   General details of dockyard administration in this chapter have been drawn from Baugh, 1977, 261–69; Haas, 1994, 7–44; Knight, 1987, xvii–lxi; MacDougall, 2012, 37–53; Morriss, 1983, 127–88.
2   MacDougall, 2012, 45.
3   Knight, 1987, xxii.
4   Haas, 1994, 76.
5   Ibid., 67–77.
6   Ibid., 76.
7   Baugh, 1977, 263.
8   Field, 1994, 8; PCL, Pescott-Frost Papers, 5, 8.
9   Morriss, 1983, 127–56; PCL, Pescott-Frost Papers, 5, 8.
10 TNA, ADM 5/793.
11 Galliver, P.W., 1987, The Portsmouth Dockyard Workforce 1880–1914, Faculty of Arts, Masters Thesis, University of Southampton, 32.
12 Ibid.
13 PCL, Pescott-Frost Papers, 5, 8.
14 Ibid.; Haas, 1994, 71–73.
15 Haas, 1994, 90.
16 PCL, Pescott-Frost Papers, 5, 8.
17 Haas, 1994, 81–90.
18 Ibid., 122–29, 174.
19 Ibid., 151–52, 172–74.
20 Ibid., 174–75, 190.
21 Galliver, op. cit., 32–46.

Chapter 5: Discipline and Security

1   All above from NMM, PFR/2/A, Chips 1674–1801.
2   NMM, PFR/2/A, Women Whipped for Theft, 1767.
3   Morriss, 1983, 94.
4   Knight, 1987, 66–71.
5   Knight, R.J.B., 1975, 'Pilfering and Theft from the Dockyards at the time of the American War of Independence', The Mariners Mirror, 61:3, 215–25.
6   Morriss, 1983, 95.
7   Knight, 1987, 69.
8   Knight, 1975, op. cit.
9   Knight, 1987, 69–71.
10 Knight, 1975, op. cit.
11 Knight, 1987, 65.
12 Webb, 1971, 7–8.
13 Field, 1994, 15.
14 History of the Ministry of Defence Police, Wikipedia, https:// en.wikipedia.org/wiki/History_of_the_Ministry_of_Defence_Police, accessed 3.1.2016.
15 NMM, PFR/2/A, Watching and Guarding the Dockyard 1678 and Onwards.
16 Ibid.
17 Ibid.
18 Morriss, 1983, 96.
19 NMM, PFR/2/A, Watching and Guarding the Dockyard 1678 and Onwards.
20 Ibid.
21 Field, 1994, 15.
22 NMM, PFR/2/A, Watching and Guarding the Dockyard 1678 and Onwards.
23 Field, 1994, 15.
24 History of the Ministry of Defence Police, op. cit.
25 Naval Dockyards Society, 2015, 118–19.
26 The Times, 13.2.1933, 'Marine Police to Replace Metropolitan Division'; History of the Ministry of Defence Police, op. cit.
27 NMM, PFR/2/A, Shipwrights etc Rioting 1743.
28 NMM, PFR/2/A, Rebellion, 1745.
29 NMM, PFR/2/A Sale of Beer to Workmen 1689–1832.

30   NMM, PFR/2/A, Beer Allowance to Smiths Working on Anchors.

31   NMM, PFR/2/A, Sale of Beer to Workmen 1689–1832.

32   NMM, PFR/2/A, Smoking 1694–97.

33   NMM, PFR/2/A, Hogs and Other Livestock, 1703–1814.

34   NMM, PFR/2/A, Workmen's Dogs, 1749.

35   Georgian London, The History of the Female Shipwright, http://georgianlondon.com/post/49247094071/the-history-of-the-female-shipwright; Mary Lacy: Inglis, L., The History of the Female Shipwrights, http://the-history-girls.blogspot.co.uk/2013/08/mary-lacy-history-of-female-shipwright.html; Wikipedia, Mary Lacy, https://en.wikipedia.org/wiki/Mary_Lacy; Mary Lacy, Alias William Chandler, http://www.kenthistoryforum.co.uk/index.php?topic=12817; all accessed 14.3.17.

36   NMM, PFR/2/A, Idleness and Impudence of Workmen, 1694.

37   NMM, PFR/2/A, Re-entry of Repentent Shipwright, 1712.

38   NMM, PFR/2/A, Workmen Absenting, 1730.

39   NMM, PFR/2/A, Employment of Dockyard Men on Private Work, 1704.

40   NMM, PFR/2/A, Punishment, 1711.

41   NMM, PFR/2/A, Payment of Wages to Dockyard Men, 1709–1805.

42   NMM, PFR/2/A, Sailmakers Irregularities, 1729.

43   NMM, PFR/2/A, Dismissal of Clerk, 1781.

44   Galliver, P.W., 1987, The Portsmouth Dockyard Workforce 1880–1914, Faculty of Arts, Masters Thesis, University of Southampton, 47.

45   Ibid., 47–51.

46   Ibid., 55–57.

47   Ibid., 56–57.

Chapter 6: Pay and Productivity

1   Haas, 1994, 26–29.

2   Ibid., 33.

3   Ibid., 29–30; Knight, 1987, xlvi–xlvii; NMM, PFR/2/A.

4   Haas, 1994, 102.

5   Ibid., 33; NMM PFR/2/A Pensions for Dockyardmen 1764 and 1771.

6   Knight, 1987, 44.

7   Haas, 1994, 34–36; Knight, 1987, xliv–xlv; NMM POR/A/26; NMM POR/D/19.

8   Knight, 1987, xxxviii.

9   Morriss, 1983, 100.

10   Ibid., 97–102, 123–26; NMM PFR/2/A, War Bonus 1801; Morriss, R., 'Labour Relations in the Royal Dockyards', 1801–1805, The Mariners Mirror, 62:4, 337–46.

11   Morriss, 1983, 102–04, 118–19.

12   Haas, 1994, 68–71.

13   Ibid., 29, 72–73.

14   TNA, ADM 3/61.

15   Haas, 1994, 3–4.

16   Ibid., 2.

17   Coats, A., 'Efficiency in Dockyard Administration 1660–1800: A Re-Assessment', in Tracy, 2002, 126–27.

18   Ibid., 129.

19   Pollard S., 'A Management Odyssey: The Royal Dockyards 1714–1914', Albion, 26 (1994), 692–93, cited by Coats, A., op. cit.,116–32.

20   Knight, R.J.B., 'From Impressment to Task Work: Strikes and Disruption in the Royal Dockyards', 1688–1788, in Day & Lunn, 1999, 13.

21   Morriss, R., 'A Management Odyssey: The Royal Dockyards 1714–1914', English Historical Review, 112/445 (1997), 226, cited by Coats, A., op. cit., 116–32.

22   Haas, 1994, 47, 65, 187.

23   Coats A., op. cit., 116–32.

24   Haas, 1994, 1–2.

25   National Audit Office, Value for Money, https://www.nao.org.uk/successful-commissioning/general-principles/value-for-money/, accessed 6 March 2017.

26   Warwick Internal Audit, https://www2.warwick.ac.uk/services/internalaudit/value/, accessed 6 March 2017.

Chapter 7: Building a Ship-of-the-Line

1   Lavery, 1991, 56.
2   Ibid., 57.
3   Baugh, 1977, 237–39.
4   Lavery, 1991, 57; Baugh, 1977, 237–39.
5   Morriss, 1983, 78–79; Goss, 1984, 29.
6   TNA, ADM 7/593.
7   Knight, 1987, 139.
8   Childs, 2007, 28–31.
9   Dodds & Moore, 1984, 17.
10   TNA ADM 7/593.
11   Knight, 1987, 138; NMM, POR/D/19, cited in Knight, 1987, 134.
12   TNA ADM 7/593.
13   Baugh, 1977, 239–40.
14   Dodds & Moore, 1984, 18.
15   Ibid., 43–44; Lavery, 1991, 67–68.
16   Lavery, 1991, 39, 67–70; Dodds & Moore, 1984, 56.
17   Lavery, 1991, 68–72; Fincham, 1825, 144–45, 175, 195, 233, 240.
18   Fincham, 1825, 3–4.
19   Dodds & Moore, 1984, 56.
20   Dodds & Moore, 1984, 58–61; Fincham, 1825, 8–10.
21   Fincham, 1825, 38.
22   Dodds & Moore, 1984, 61–64; Lavery, 1991, 80–82; Fincham, 1825, 9–19, 21–22.
23   Fincham, 1825, 28–29, 35–37; Dodds & Moore, 1984, 64–74; Lavery, 1991, 83–97.
24   Lavery, 1991, 98–111; Dodds & Moore, 1984, 84–85.
25   Seppings, R., 1818, 'On the great strength given to Ships of War by the application of Diagonal Braces', *Philosophical Transactions of the Royal Society of London*, London, 1–8.
26   Lavery, 1991, 117–28; Dodds & Moore, 1984, 89–102; Fincham, 1825, 5–7, 77–78, 83–84, 115.
27   Lavery, 1991, 129–39; Dodds & Moore, 1984, 99, 102, 105.
28   Lavery, 1991, 140; Dodds & Moore, 1984, 109–10.
29   Dodds & Moore, 1984, 105; Lavery, 1991, 141–48.
30   Dodds & Moore, 1984, 115, 122; Lavery, 1991, 156.
31   Patterson, 2008, 42.
32   Globalsecurity.org, Sailing Ship Hull Timbers, http://www.globalsecurity.org/military/systems/ship/sail-hull-timbers.htm, accessed 6.7.2017.

Chapter 8: The Royal Naval Academy

1   Coad, 1989, 75.
2   Dickinson, H.W., 2003, 'The Portsmouth Naval Academy, 1733–1806', *The Mariner's Mirror*, 89:1, 17–30.
3   Coad, 1989, 75.
4   Coad, 2013, 382
5   Ibid.; British Listed Buildings, Former Naval Academy, http://www.britishlistedbuildings.co.uk/en-476657-former-royal-naval-academy-buildings-num#.VsRHG-YaaJc, accessed 19.2.16; Coats, A. et al , 2015, 120.
6   Dickinson, op. cit., 17.
7   Baugh, 1977, 59.
8   NMM, PFR/2/A, Royal Academy 1729 and onwards.
9   Baugh, 1977, 58–62.
10   Dickinson, op. cit., 19.
11   Ibid., 20–21, 26.
12   Rodger, 2004, 510–11.
13   Dickinson, op. cit., 21–22.
14   Ibid.
15   NMM, PFR/2/A, Family Squabble 1766.
16   Ibid.
17   Dickinson, op. cit., 23–24, 27.
18   NMM, POR/F/15, Commissioner Gambier to Navy Board.
19   Dickinson, op. cit., 25–26.
20   Ibid., 24.
21   NMM, PFR/2/A, Royal Academy Punishments etc 1735–1777.
22   NMM, PFR/2/A, Money Lending etc, 1757.
23   Lloyd, C., 1966, 'The Royal Naval Colleges at Portsmouth and Greenwich', *The Mariner's Mirror*, 52:1, 145–56.
24   NMM, POR/C/22, Commissioner Gambier to Secretary to Admiralty.
25   Knight, 1987, lxii, ref 61.

26 NMM, PFR/2/A, Royal Academy Punishments 1735–1777.

27 NMM, POR/C/22, Commissioner Gambier to Secretary to the Admiralty; PFR/2/A, Royal Academy Punishments 1735–1777.

28 NMM, PFR/2/A, Royal Academy Fatality, 1748.

29 Dickinson, op. cit., 18, 27.

30 Coad, 2013, 382.

31 Lloyd, op. cit., 146–47.

32 Ibid., 147.

33 Ibid.

34 Coad, 2013, 382.

35 Coad, 1989, 76.

Chapter 9: The Victorian Dockyard

1   Archibald, 1972, 155–57.

2   Evans, 2004, 32–35, 65.

3   Ibid., 26–29.

4   Laing, 1985, 12.

5   Ibid., 13–14; Winton, 1989, 103.

6   Coats, A. et al, 2015, 3.

7   Evans, 2004, 39.

8   Laing, 1985, 18.

9   The Times, 16.5.1848.

10 Riley, 1985, 15; Patterson, 1989, 2–3.

11 Goss, 1984, 31.

12 Patterson, 1989, 4, 57.

13 Riley, 1985, 15.

14 Wells, 1987, 138.

15 Riley, 1985, 16.

16 Coats, A. et al, 2015, 137–38.

17 Riley, 1985, 16; Coad, 2013, 193–94.

18 Riley, 1985, 16; Coad, 2013, 194.

19 Coad, 2013, 195.

20 Riley, 1985, 24.

21 Coats, A. et al, 2015, 162.

22 Riley, 1985, 16; Coad, 2013, 193–94.

23 Riley, 1985, 16; Coats, A. et al, 2015, 126.

24 Evans, 2004, 49–50;Coats, A. et al, 2015, 133.

25 Riley, 1985, 25.

26 The News, Portsmouth, 2.1.1991.

27 Riley, 1985, 21.

28 The News, Portsmouth, 2.1.1991.

29 Riley, 1985, 21, 25.

30 The News, Portsmouth, 2.1.1991; Riley, 1985, 25.

31 Hawkins, D., Butler, C., and Skelton, A., 'Iron Slip Cover Roofs of the Royal Dockyards 1844–1857', Transactions of the Naval Dockyard Society, 9, 65–82; Riley, 1985, 21.

32 The Times, 1.8.1855.

33 The Marlborough was hulked in 1904 at Portsmouth as part of the HMS Vernon establishment.

34 The Times, 27.4.1857.

35 TNA, ADM 135/409, f14.

36 TNA, ADM 135/409, f14, Report on trial 12 August 1858.

37 TNA, ADM 135/409, f1, 2, 4, 10, reports on trials 1864–66.

38 TNA, ADM 1/5864, f A1513, 15 Oct 1864, Sherard Osborn, Capt., Royal Sovereign to C-in-C.

39 Winton, 1989, 104; Goss, 1984, 34–35.

40 TNA, ADM 1/5943, Captain, Excellent, to Commander-in-Chief.

41 Prince of Wales was renamed Britannia in 1869 when she took over the officer cadet training role at Dartmouth, replacing the Britannia of 1820.

42 On 19 October 1876 she was renamed Worcester, to take on a role as a training ship at Greenhithe for the Thames Nautical Training College. She remained in this role until her sale in July 1948, and foundered in the River Thames on 30 August 1948, but was raised in May 1953 and broken up.

43 The Times, 14.11.1859.

44 TNA, ADM 135/492, f14, Report on trial 5 July 1860.

45 Hendriks, 2014, 5–18.

46 Ibid., 13–18.

47 Ibid., 18–21; The Times, 21.4.1860.

48 Preston & Major, 2007, 34.

49 Ibid., 156, 190; TNA, ADM 87/70, fS1719.

50 The Times, 9.9.1872.

51 Hendriks, 2014, 42–43.

52 Coad, 2013, 199–202; Hendriks, 2014, 43–44.

53 Coad, 2013, 199–202; Hendriks, 2014, 51–76.
54 Goss, 1984, 40–45.
55 NMM, POR/P/77; NMM, POR/P/81.
56 *The Times*, 13.7.1871.
57 Haas, 1994, 98, 114–15.
58 Ibid., 115, 129.
59 Galliver, P.W., 1987, *The Portsmouth Dockyard Workforce 1880–1914*, Faculty of Arts, Masters Thesis, University of Southampton, 1–9, 25.
60 Ibid., 26–30.
61 Haas, 1994, 165–66.
62 Riley, 1985, 22–23; Coad, 2013, 184.
63 Coats, A. et al, 2015, 163.
64 Evans, 2004, 194–95.
65 Riley, 1985, 22–23; Patterson, 1989, 12; Leather, 2005, 11.
66 Riley, 1985, 22–23; Patterson, 1989, 12–22; Coad, 2013, 40–41; Leather, 2005, 130.
67 Patterson, 1989, 14–16.
68 Coad, 2013, 40–41.
69 Riley, 1985, 24–25.
70 Coats, A. et al, 2015, 167.
71 *The Times*, 28.4.1876; *Hampshire Telegraph*, 28.4.1876.
72 Goss, 1984, 46, 50; Winton, 1989, 107.
73 Patterson, 1989, 28–29; Dittmar & Colledge, 1972, 30.
74 Hamilton, 2005, 396–97.
75 Patterson, 1989, 59.
76 PP, Health of the Navy Returns, 1908, Lxv Cd 296, 319.
77 Galliver, op. cit., 18–19.
78 Haas, 1994, 148.
79 PP, Health of the Navy Returns, 1908, LXV Cd 296, 319.
80 Haas, 1994, 158–62.
81 Goss, 1984, 58–69.
82 *The Times*, 27.2.1891.
83 Goss, 1984, 61–69.
84 Galliver, op. cit., 296; Riley, 1976, 9.
85 *The Times*, 27.6.1897.
86 Winton, 1989, 117–18.
87 Yates, R.W., 1962, 'From Wooden Walls to Dreadnoughts in a Lifetime', *The Mariner's Mirror*, 48:4, 291–303.

Chapter 10: The Dockyard Apprentice

1 Macleod, N., 1925, 'The Shipwrights of the Royal Dockyards', *The Mariner's Mirror*, 11:3, 281–87.
2 Haas, 1994, 58–59; Knight, R.J.B., 'From Impressment to Task Work: Strikes and Disruption in the Royal Dockyards', 1688–1788, in Day & Lunn, 1999, 7.
3 Haas, 1994, 58–59; Macleod, op. cit., 287.
4 Board of Education, 1916, The Admiralty Method of Training Dockyard Apprentices, Educational Pamphlet No. 32, HMSO.
5 Ibid.
6 Casey, N., 'Class Rule: The Hegemonic Role of the Royal Dockyard Schools, 1840–1914', in Day & Lunn, 1999, 66.
7 Ibid., 68.
8 Ibid., 69.
9 Board of Education, op. cit.; Thomas, K., 2009, Portsmouth Dockyard School, PRDHT, http://portsmouthdockyard.org.uk/images/uploads/documents/PORTSMOUTH_DOCKYARD_SCHOOL.pdf, accessed 18.3.2017.
10 Benyon, P., Entry of Apprentices in Her Majesty's Dockyards, http://www.pbenyon.plus.com/Navy_List_1879/Dockyd_Apps.html, accessed 11.11.15.
11 Casey, op. cit., 69–70.
12 Macleod, op. cit., 286.
13 Casey, op. cit., 81.
14 Thomas, op. cit.
15 King, F.E., 2013, 'The Paucity of Shipwrights in Royal Naval Dockyards during the Second World War', *The Mariner's Mirror*, 99:2, 228–33, citing ADM 116/4722.
16 Ibid.
17 King, F.E., 2013, 'The Royal Dockyard Schools and their Education System', *The Mariner's Mirror*, 99:4, citing ADM 116/4722.
18 PRDHT, Royal Dockyard Apprenticeships, http://portsmouthdockyard.org.uk/version5/about/exhibitions, accessed 18.3.17
19 Thomas, op. cit.
20 Ibid.
21 Lane,1988, 5, 15–36, 42.
22 Lunn & Day, 1998, 5.

23   Lunn & Day, 1998, 6; Day & Pritchard, 1999, 14.

24   Ibid., 13–14.

25   Lunn & Day, 1998, 5.

26   Day & Pritchard, 1999, 14.

27   Ibid., 17.

28   NMRN, 1992.235, Dockyard Days.

29   Lunn & Day, 1998, 6–7.

30   Ibid., 8.

31   Day & Pritchard, 1999, 17.

32   NMRN, 1992.235, Dockyard Days.

33   Lunn & Day, 1998, 9–10.

34   Day & Pritchard, 1999, 17.

35   Ibid., 22.

36   Lunn & Day, 1998, 10.

37   Day & Pritchard, 1999, 19–20.

38   Ibid., 20–21.

Chapter 11: The Edwardian Era

1   Brown, D.K., 1997, 180–82.

2   Brown, D.K., 1997, 181–84.

3   Dittmar & Colledge, 1972, 32–36.

4   Patterson, 1989, 32.

5   Ibid., 33.

6   Patterson, 1984, 11–12.

7   Patterson, 1989, 39.

8   Riley, 1985, 23.

9   Engineering, 13 February 1914, 205.

10   Coats, A. et al, 2015, 170.

11   Patterson, 1984, 12.

12   Ibid., 16–17.

13   Patterson, 1989, 38.

14   'Floating Docks', The Invergordon Archive, http://www.
     theinvergordonarchive.org/groups.asp?id=42, accessed 12.2.14.

15   The Times, 22.12.1913, 'Dockyard Fire at Portsmouth'; Bob
     Hind, The News, Portsmouth, 3.5.14; A Thousand Men
     Fought Dockyard Blaze, http://www.portsmouth.co.uk/
     heritage/a-thousand-men-fought-portsmouth-dockyard-blaze-1-
     6034599#ixzz47sjyLGrH), accessed 15.6.16.

16   Haas, 170–171.

17   Ibid., 158.

18   Ibid., 172.

19   Ibid, 172–75.

20   TNA, ADM181/158, Navy Estimates, Miscellaneous Papers
     1915–17.

21   Thomas, R., 'The Building of HMS Dreadnought and Dreadnought
     Battlecruiser Gunnery', Transactions of the Naval Dockyard
     Society, 3, 39.

22   Brown, D.K., 1979, 'The Design and Construction of the Battleship
     Dreadnought', Warship, 13, 52.

23   NMRN, 252/12/34, Repairs to the Fleet.

24   NMRN, 252/12/21, Notes for a Meeting of C-in-C Home Ports
     and Admirals Superintendent, Defects and Repairs.

25   NMRN, 252/12/20, Repairs to the Fleet.

26   Bacon, 1929, 12–20.

27   NMRN, 253/17, Naval Establishments Enquiry Committee,
     Preliminary Visit to Naval Establishments from 1st to 12th May
     1905.

28   Fisher, 1919, 57–58.

29   Hansard, 29 March 1909, vol. 3 c10.

30   Derived from data in TNA, ADM181/159 and ADM181/158,
     Navy Estimates: Programme of Works for the Home
     Dockyards and Naval Yards Abroad, 1913–14 and 1905–06
     respectively.

31   Derived from data in MeasuringWorth.com, The Annual RPI and
     Average Earnings for Britain, 1209 to 2012 (New Series), www.
     measuringworth.com/ukearncpi/, accessed 14.2.14.

32   Derived from data in Dittmar & Colledge, 1972.

33   Derived from data in TNA, ADM181/159.

34   Hythe, 1913, 5.

35   Ibid., 4, 7.

36   Ibid., 5–6.

37   Derived from data in I. Sturton (ed.), Conway's Battleships,
     Conway, London, 2008, 101–16.

38   Derived from data in TNA, ADM181/159 and ADM181/158.

39   Ibid.

Chapter 12: Building *Dreadnought*

1   Brown, D.K., 1979, 'The Design and Construction of the Battleship *Dreadnought*', *Warship*, 13, 39–49.
2   Thomas, R.D., 2007, 'The Building of HMS Dreadnought and Dreadnought Battlecruisers 1905–1916', *Transactions of the Naval Dockyards Society*, 3, 37–38, 45–46.
3   Thomas & Patterson, 1998, 7.
4   Brown, D.K., 1979, op. cit., 48–51; Brown, D.K., 1997, 190.
5   Brown, D.K., 1979, op. cit., 52, citing Narbeth, J.H., 1922, 'Three Steps in Naval Construction', *Transactions of the Institution of Naval Architects*.
6   *The Times*, 2.4.1919, Obituary.
7   Brown, D.K., 1979, op. cit., 51–52; Brown, 1997, 190.
8   Brown, D.K., 1979, op. cit., 52.
9   Patterson, B., 2006, 'Building of Dreadnought', lecture delivered to a conference at the Royal Naval Museum, Portsmouth, 3.6.06; Brown, D.K., 1979, op. cit., 40–42.
10 Thomas & Patterson, 1998, 1–115; Johnston & Buxton, 2013, 108–35.
11 Thomas & Patterson, 1998, 52–73; Johnston & Buxton, 2013, 108–35.
12 *The Times*, 12.2.1906, 'The King and the Dreadnought'.
13 Patterson, 1984, 12.
14 Johnston & Buxton, 2013, 148.
15 Roberts, 2013, 24, 27.
16 Ibid., 15–18.
17 Johnston & Buxton, 2013, 237.
18 Roberts, 2013, 18.
19 Data drawn from Sturton, 2008.
20 Dittmar & Colledge, 1972, 33–34.

Chapter 13: The First World War

1   TNA, ADM1/8387/223, Labour Requirements in HM Dockyards, July–September 1914.
2   TNA, T11803, Revision of Pay in Dockyards, 1915.
3   Courtney and Patterson, 2005, 28.
4   Ibid.
5   TNA, T11803.
6   Galliver, P., 'Trade Unionism in Portsmouth Dockyard', in Lunn & Day, 1999, 102–03.
7   Ibid., 120.
8   TNA, T11803.
9   *The Times*, 27.3.1917.
10 TNA, ADM1/8560/157, Dockyard Employees: Papers Relating to Various Arbitration Awards, 1919.
11 TNA, T1/12253, The Admiralty's Grant of Pay Increases to Female Dockyard Employees without Treasury Approval, 1918.
12 Day and Pritchard, 1999, 33.
13 TNA, ADM181/100.
14 NMRN, Navy Estimates, Vote 8, 1919–20.
15 Derived from data in NMRN, Navy Estimates, Vote 8, 1914–15; TNA, ADM181/100; Lunn, K., and Day, A., 'Labour Relations in the Royal Dockyards', in Lunn & Day, 1999, 129.
16 MeasuringWorth.com, The Annual RPI and Average Earnings for Britain, 1209 to 2012 (New Series), www.measuringworth.com/ukearncpi/, accessed 14.2.14.
17 Galliver, P., 'Trade Unionism in Portsmouth Dockyard', in Lunn & Day, 1999, 122; TNA, ADM179/69, f593.
18 TNA, ADM179/69, f593.
19 Ibid., f594.
20 Patterson, 1989, 41.
21 Haas, 1994, 177.
22 Patterson, 1989, 42.
23 TNA, ADM181/159.
24 NMRN, Dockyard Expense and Manufacturing Accounts 1913–14.
25 TNA, ADM179/75, Portsmouth General Orders, Docking, 1914.
26 TNA, ADM179/69, f593.
27 Courtney & Patterson, 2005, 28.
28 1914 Guide for Visitors to the Dockyard, http://www.devonportonline.co.uk/historic_devonport/navy-dockyard/dockyard/1914-guide-dockyard.aspx, accessed 27.2.14, reproduced from Ward Lock, A Pictorial and Descriptive Guide to Plymouth, Stonehouse and Devonport with Excursions by River, Road and Sea, Ed. 5 rev, Ward Lock, London, 1914.
29 TNA, ADM179/69, f593.

30   Sadden, 2012, 126.

31   TNA, ADM179/69, f587.

32   Lunn & Day, 1999, 135–37.

33   Sadden, 2012, 127.

34   Gates, 1919, 57–58.

35   Clark, C., 2014, 'Women at Work in Portsmouth Dockyard 1914–19', a lecture delivered to Naval Dockyards Society conference, Greenwich, March 2014.

36   Lunn & Day, 1999, 135–37.

37   TNA, ADM 179/69, Portsmouth dockyard records, 1916–24, f563.

38   Ibid., f552–594.

39   *The Times*, 17.10.1913.

40   Lane, 1998, 25.

41   TNA, ADM 179/69, Portsmouth dockyard records, 1916–24, f563.

42   http://www.naval-history.net/OWShips-WW1-02-HMS_Princess Royal.htm, accessed 1.6.16.

43   TNA, ADM 179/69, Portsmouth dockyard records, 1916–24, f563.

44   Ibid.

45   Dittmar & Colledge, 1972, 136.

46   Courtney, S., and Patterson, B., *Home of the Fleet*, 31.

47   TNA, ADM 179/69, f556–557.

48   Ibid., f568–569.

49   Ibid., f557.

50   Ibid., f558–559.

51   Ibid., f559–561.

52   Ibid., f561, 569–570.

53   Ibid., f594.

54   Sadden, 2011, 44.

55   *The Times*, 12.1.1918.

Chapter 14: The Inter-War Years

1   Washington Naval Treaty, https://en.wikipedia.org/wiki/ Washington_Naval_Treaty, accessed 5.6.16.

2   London Naval Treaty, https://en.wikipedia.org/wiki/London_ Naval_Treaty, accessed 5.6.16.

3   TNA, ADM179/69 f593; Hansard, 1946.

4   Peden, G., 2000, *The Treasury and British Public Policy: 1906–1959*, Oxford, Oxford University Press, 169.

5   Patterson, 1984, 16–17.

6   Ibid., 11, 13.

7   PRDHT, 1923 – 250-ton Floating Crane, http:// portsmouthdockyard.org.uk/timeline/details/1923-250-ton- floating-crane, accessed 1.2.14.

8   Ibid.

9   Patterson, 1984, 15.

10   Winton, 1989, 137–38; Courtney & Patterson, 2005, 37–39; Goss, 1984, 86; Lenton, 1998, 44.

11   Helder Line, http://www.helderline.nl/tanker/772/ murex+%282%29/, accessed 4.7.17.

12   Day and Pritchard, 1999, 33.

13   TNA, ADM 1/8636/36.

14   Hansard, 1946; Courtney & Patterson, 2005, 37–38; Winton, 1989, 137.

15   Courtney & Patterson, 2005, 40.

16   http://www.naval-history.net/xGM-Chrono-02BC-HMS_Renown. htm, http://www.naval-history.net/xGM-Chrono-02BC-HMS_ Repulse.htm, accessed 2.2.14.

17   Courtney & Patterson, 2005, 41.

18   Winton, 1989, 139; Patterson, 1984, 7.

19   Courtney & Patterson, 2005, 47–48.

20   Patterson, 1984, 3–4; Courtney & Patterson, 2005, 41–42.

21   Patterson, 1984, 9–10, 14.

22   Coats, A. et al, 2015, 107.

23   *The Times*, 16.8.1927.

24   *The Times*, 7.8.1928, 9.8.1928, 18.8.1928, 20.8.1928.

25   *The Times*, 8.8.38.

26   *The Times*, 29.7.36, 31.7.36, 1.8.36.

27   *The Times*, 10.3.39.

28   Introduction of paid leave, PRDHT, http://portsmouthdockyard. org.uk/timeline/time/1914-1984 , accessed 5.6.16.

29   Courtney & Patterson, 2005, 49–50; Lenton, 1998, 155.

30   Hannan, 1987, 130.

31   Goss, 1984, 90, 92.

32   Sturton, 2008, 109.

33   http://www.naval-history.net/xGM-Chrono-01BB-HMS_Queen_
      Elizabeth.htm, accessed 6.6.16.

34   Sturton, 2008, 122.

35   http://www.naval-history.net/xGM-Chrono-02BC-HMS_Renown.
      htm, accessed 6.6.16.

36   Hansard, 1946.

37   Goss, 94.

38   Courtney & Patterson, 2005, 55.

Chapter 15: The Second World War

1   Courtney & Patterson, 2005, 70; http://www.naval-history.net/
      xGM-Chrono-01BB-HMS_Queen_Elizabeth.htm, http://www.
      naval-history.net/xGM-Chrono-01BB-HMS_Nelson.htm, http://
      www.naval-history.net/xGM-Chrono-01BB-HMS_Resolution.htm,
      accessed 11.9.15.

2   Courtney & Patterson, 2005, 71.

3   The Times, 26.6.1939, 21.7.1939.

4   Coats, A. et al, 2015, 186.

5   Courtney & Patterson, 2005, 70–71.

6   Coats, A. et al, 2015, 114.

7   Patterson, 2008, 114.

8   Day, A., 1998, The forgotten 'mateys': women workers in
      Portsmouth Dockyard, England, 1939–45, Women's History
      Review, 7:3, 361–82.

9   Hansard, 16 April 1946.

10   Hansard, 1946.

11   Keel, A., Eighty Six to One, http://portsmouthdockyard.org.uk/
      personal-accounts, accessed 12.9.16.

12   Anon, 1942–1944 My Career Begins, http://portsmouthdockyard.
      org.uk/personal-accounts, accessed 12.9.16.

13   Coats, A. et al, 2015, 111.

14   1943 Frederick Street, Gloucester Street and Marlborough Row,
      PRDHT, http://portsmouthdockyard.org.uk/timeline/details/1943-
      frederick-street-gloucester-street-marlborough-row, accessed
      12.9.16.

15   Coats, A. et al, 2015, 15–17; The Portsmouth Blitz, http://www.
      welcometoportsmouth.co.uk/the%20blitz.html, accessed 11.9.16.

16   Courtney and Patterson, 2005, 66–69.

17   TNA, ADM 1/10949; Battle of Britain Historical Society, The Battle
      of Britain 1940, http://www.battleofbritain1940.net/0024.html.

18   TNA, ADM 1/10949; James, 1946, 79.

19   Courtney & Patterson, 2005, 70; HMS Acheron destroyer,
      NavalHistory.net, http://www.naval-history.net/xGM-Chrono-
      10DD-13A-HMS_Acheron.htm; HMS Bulldog destroyer, Naval.
      History.net, http://www.naval-history.net/xGM-Chrono-10DD-
      15B-HMS_Bulldog.htm; accessed 11.6.17.

20   TNA, ADM 267/130.

21   James, 1946, 95–96.

22   The Times, 1.2.1941.

23   Mc Gowan, 1999, 40.

24   TNA, ADM 267/130.

25   James, 1946, 111–112.

26   TNA, ADM 267/130.

27   Patterson, 2008, 106, 108.

28   TNA, ADM 179/171.

29   Ibid.

30   World War II Today, June 22 1943 Heroic Rescue on Burning
      HMNZS Achilles, http://ww2today.com/22nd-june-1943-heroic-
      rescue-on-burning-hmnzs-achilles, accessed 18.7.17.

31   D-Day exploding the myths of the Normandy landings, CNN.com,
      http://edition.cnn.com/2014/06/05/opinion/opinion-d-day-myth-
      reality/index.html; D-Day Museum, http://www.ddaymuseum.
      co.uk/d-day/d-day-and-the-battle-of-normandy-your-questions-
      answered#stand, accessed 22.9.15.

32   TNA, ADM 1/13186; Winton, 1989, 160.

33   D-Day Museum, http://www.ddaymuseum.co.uk/d-
      dayonyourdoorstep/details/mulberry-harbour-construction-site-
      hayling-island, accessed 22.9.15.

34   Doughty, 1994, 40–43, 113–14; Winton, 1989, 161–62.

35   Courtney & Patterson, 2005, 74–75; TNA, ADM179/415.

36   TNA, ADM 179/415; Patterson, 2008, 119–20.

37 TNA, ADM 179/415.

38 Ibid.

39 HMS *Roberts* monitor, NavalHistory.net, http://www.naval-history. net/xGM-Chrono-03Mon-HMS_Roberts.htm, accessed 10.9.15; HMS *Arethusa* cruiser, NavalHistory.net, http://www. naval-history.net/xGM-Chrono-06CL-HMS_Arethusa.htm, accessed 10.9.15.

40 Lunn & Day, 1998, 14.

41 NMRN, Mss 1992.235, D.K. Muir, Dockyard Days.

42 Lunn & Day, 1998, 14.

43 Ibid., 23–24.

44 Ibid., 26.

Chapter 16: The Cold War

1 Bush, 2011, 217–34.

2 Coats, A. et al, 2015, 22; Hansard Debates, 1.7.1963.

3 Coats, A. et al, 2015, 20–26.

4 Blackman, R. (ed.), 1952, *Jane's Fighting Ships 1952–53*, London, Samson, Low, Marston; Moore, J.E. (ed.), 1982, *Jane's Fighting Ships 1982–83*, London, Jane's.

5 NMRN, 1984.610.

6 HMS *Relentless* destroyer, Naval History.net, http://www. naval-history.net/xGM-Chrono-10DD-53R-HMS_Relentless.htm, accessed 1.2.15.

7 HMS *Verulam* destroyer, NavalHistory.net, http://www. naval-history.net/xGM-Chrono-10DD-61V-HMS_Verulam.htm, accessed 1.2.15; HMS *Troubridge* destroyer, NavalHistory.net, http://www.naval-history.net/xGM-Chrono-10DD-56T-HMS_ Troubridge.htm, accessed 1.2.15.

8 Brown, D.K., and Moore, 2003, 41–42; Hobbs, 2013, 267–70.

9 Courtney & Patterson, 2005, 91, 105.

10 Ibid., 90; Patterson, 2008, 116.

11 Hansard, 4 May 1960, vol. 622, cc1055-6.

12 Hobbs, 2013, 193, 246, 248; Boniface, 2007, 120.

13 McCart, 1999, 128.

14 Courtney & Patterson, 2005, 105.

15 Ibid., 115–16.

16 Boniface, 2007, 176–78; Patterson, 1991, 36.

17 Courtney & Patterson, 2005, 120.

18 Navy Days 1961, Portsmouth: Navy Days Secretary.

19 Hannan, 1987, 4; RFA Nostalgia, RMAS, http://rfanostalgia.org/ gallery3/index.php/RMAS, accessed 13.6.17.

20 Coats, A. et al, 2015, 187.

21 Ibid., 38.

22 Patterson, 2008, 117.

23 Coats, A. et al, 2015, 133.

24 Ibid., 38.

25 Clark, C., 2008, 'Adaptive re-use and the Georgian storehouses of Portsmouth – naval storage to museum', *Transactions of the Naval Dockyard Society*, 4:32; Lily Lambert McCarthy, http:// www.telegraph.co.uk/news/obituaries/1516352/Lily-Lambert-McCarthy.html, accessed 13.6.17.

26 Patterson, 2008, 118.

27 Coats, A. et al, 2015, 38.

28 Patterson, 2008, 105.

29 Coats, A. et al, 2015, 38.

30 Patterson, 2008, 174.

31 Ibid., 118.

32 Coats, A. et al, 2015, 175.

33 *The News*, Portsmouth, 2.1.1991.

34 Patterson, 2008, 117–18; Coats, A. et al, 2015, 173.

35 Courtney & Patterson, 2005, 118; Coats, A. et al, 2005, 156.

36 Patterson, 2008, 119.

37 Coats, A. et al, 2005, 145.

38 Courtney & Patterson, 2005, 119–20.

Chapter 17: The Rundown of the Dockyard

1 NMRN, Mss 1984.610.

2 Courtney & Patterson, 2005, 132.

3 *The Times*, 10.9.1981.

4 NMRN, Mss 1984.610.

5 Patterson, 1998, 45–47, 62.

6 NMRN, Mss 1984.610.

7 Ibid.

8 Coats, A. et al, 2015, 177.

9 NMRN, Mss 1984.610.

10 Day & Pritchard, 1999, 7, 10.

11 David Barber, PRDHT, http://portsmouthdockyard.org.uk/foundry story.pdf, 65–66, accessed 16.6.17.

12 *The News*, Portsmouth, Dockyard 500 supplement, 12.6.1995.

13 *The News*, Portsmouth, 19.6.1982.

14 Day & Pritchard, 1999, 11.

15 *The News*, Portsmouth, 28.4.1982.

16 Day & Pritchard, 1999, 37.

17 *The Times*, 1.10.84.

Chapter 18: Portsmouth Naval Base and the Historic Dockyard

1 Courtney & Patterson, 2005, 140, 150.

2 Day & Pritchard, 1999, 12.

3 *The News*, Portsmouth, 13.10.1998.

4 *The News*, Portsmouth, 31.10.2004.

5 Coats, A. et al, 2015, 38.

6 Socio-Economic Impact Assessment of Portsmouth Naval Base, University of Portsmouth, 2007, https://www.economicsnetwork.ac.uk/sites/default/files/Dave%20Clark/PortsmouthNavalBaseFullReport.pdf, accessed 6.3.15.

7 Coats, A. et al, 2015, 105.

8 Ibid., 139.

9 *The News*, Portsmouth, 24.4.2003.

10 Horne, R.S., 1966, *Her Majesty's Dockyard at Portsmouth*, Portsmouth: Ministry of Public Works; NMM, PFR/2/A, Timber for Dockyard (1695, 1697).

11 *The Times*, 27.2.1915; the statue is now situated between the Porter's Lodge and No.5 Boathouse, but was formerly opposite No. 11 Storehouse.

Appendix: Ships Built in Portsmouth Royal Dockyard

1 Before 1873 the tonnage is the builder's measurement, a capacity measurement arrived at from, perhaps, the fifteenth century, by calculating the number of tuns (casks) of wine that the ship could carry. From the early eighteenth century it was calculated using the formula (k x b x ½b)/94, where k is keel length and b the beam outside the planking (but inside the wales). From 1873 displacement tonnage is used, changed in 1926 to standard displacement.

2 *Medina* was theoretically a rebuild of the 1703 yacht *Portsmouth*, but was a new vessel.

3 The last sailing ship without engines to be built in the dockyard.

4 *Princess Royal* was launched as *Prince Albert* but was renamed on 26 June 1853.

5 Built at Haslar Gunboat Yard.

6 Built at Haslar Gunboat Yard.

7 Built at Haslar Gunboat Yard.

8 Built at Haslar Gunboat Yard.

9 Built at Haslar Gunboat Yard.

10 Built at Haslar Gunboat Yard.

11 The last all wooden-hulled ship to be built in the dockyard.

12 The first iron-hulled ship to be built in the dockyard.

13 Incomplete at the end of the war and scrapped 1946.

14 Incomplete at the end of the war and scrapped 1947.

BIBLIOGRAPHY

Archibald, E., 1972, *The Wooden Fighting Ship*, Poole: Blandford.

Bacon, R.H., 1929, *The Life of Lord Fisher of Kilverstone, vol. 2*, London: Hodder and Stoughton.

Baugh, D.A. (ed.), 1977, *Naval Administration 17151750*, London: Navy Records Society.

Boniface, P., 2007, *Battle Class Destroyers*, Liskeard: Maritime.

Brown, D., 1987, *The Royal Navy and the Falklands War*, London: Lee Cooper.

Brown, D.K., 1997, *Warrior to Dreadnought*, London: Chatham.

Brown, D.K., 1999, *The Grand Fleet*, London: Chatham.

Brown, D.K., and Moore, G., 2003, *Rebuilding the Royal Navy*, London: Chatham.

Brown P.J., 2016, *Maritime Portsmouth*, Stroud: History Press.

Burton, L. (ed.), 1986, *Attentive to Our Duty*, Gosport: Gosport Society.

Bush, S., 2011, *British Warships and Auxiliaries 1952*, Liskeard: Maritime.

Childs, D., 2007, *The Warship Mary Rose*, London: Chatham.

Coad, J., 1981, *Historic Architecture of HM Naval Base Portsmouth, 1700–1850*, Portsmouth: Portsmouth Royal Naval Museum.

Coad, J., 1983, *Historic Architecture of the Royal Navy*, London: Gollancz.

Coad, J., 1989, *The Royal Dockyards 1690–1850*, Aldershot: Scolar.

Coad, J., 2005, *The Portsmouth Block Mills*, Swindon: English Heritage.

Coad, J., 2013, *Support for the Fleet: Architecture and Engineering of the Royal Navy's Bases 1700–1914*, Swindon: English Heritage.

Coats, A., Davies, J.D., Evans, D., and Riley, R., 2015, *20th Century Naval Dockyards: Devonport and Portsmouth Characterisation Report*, Portsmouth: The Naval Dockyards Society.

Colledge, J.J., and Warlow, B., 2006, *Ships of the Royal Navy*, London: Chatham.

Courtney S., and Patterson, B., 2005, *Home of the Fleet*, Stroud: Sutton.

Day, A., and Pritchard, G. (eds), 1999, *Staying Afloat: Recollections of Portsmouth Dockyard 1950–present day*, Portsmouth: University of Portsmouth.

Derrick, C., 1806, *Memoirs of the Rise and Progress of the Royal Navy*, London: C. Derrick (of the Navy Office).

Dittmar, F.J., and Colledge, J.J., 1972, *British Warships 1914–1919*, Shepperton: Ian Allan.

Dodds, J., and Moore, J., 1984, *Building the Wooden Fighting Ship*, New York: Facts on File.

Doughty, M. (ed.), 1994, *Hampshire and D-Day*, Crediton: Hampshire/Southgate.

Drummond, M., and McInnes, R. (eds), 2001, *The Book of the Solent*, Chale: Cross.

Evans, D., 2004, *Building the Steam Navy*, London: Conway.

Field, J., 1994, 'Portsmouth Dockyard and its Workers 1815–1875', *Portsmouth Papers*, No. 64, Portsmouth: City of Portsmouth.

Fincham, J., 1825, *An Introductory Outline of the Practice of Shipbuilding, Etc*, Portsea.

Fisher, J., 1919, *Records*, London: Hodder and Stoughton.

Gates, W., 1919, *Portsmouth and the Great War*, Portsmouth: Evening News.

Goss, J., 1984, *Portsmouth-built Warships, 1497–1967*, Emsworth: Kenneth Mason.

Haas, J.M., 1994, *A Management Odyssey: The Royal Dockyards, 1714–1914*, Maryland: University Press of America.

Hamilton, C.I., 2005, *Portsmouth Dockyard Papers 1852–1869: From Wood to Iron*, Portsmouth Record Series, Winchester: Hampshire County Council.

Hannan, B., 1987, *Fifty Years of Naval Tugs*, Liskeard: Maritime Books.

Hendriks, S.P.C., 2014, *Haslar Gunboat Yard, Gosport, Hampshire*, Portsmouth: English Heritage.

Hobbs, D., 2013, *British Aircraft Carriers*, Barnsley: Seaforth.

Horne, R.S., 1966, *Her Majesty's Dockyard at Portsmouth*, Portsmouth: Ministry of Public Works.

Hythe, Viscount, 1913 (ed.), *The Naval Annual*, Portsmouth: J. Griffin.

James, W., 1946, *The Portsmouth Letters*, London: Macmillan.

Jane, F.T., 1912, *The British Battle-Fleet*, London: Partridge.

Johnston, I., and Buxton, I., 2013, *The Battleship Builders*, Barnsley: Seaforth.

Knight, R.J.B., 1987, *Portsmouth Dockyard Papers 1774–1783: The American War*, Portsmouth: City of Portsmouth.

Laing, E.A.M., 1985, 'Steam Wooden Warship Building in Portsmouth Dockyard 1832–52', *Portsmouth Papers*, No. 42, Portsmouth: City of Portsmouth.

Lane, E., 1988, *The Reminiscences of a Portsmouth Dockyard Shipwright 1901–1945*, Ryde: K., T., R. and G. Lane.

Lavery, B., 1991, *Building the Wooden Walls*, London: Conway.

Leather, D., 2005, *Contractor Leather*, Ilkley: Leather Family History Society.

Lenton, H.T., 1998, *British and Empire Warships of the Second World War*, London: Greenhill.

Lewis, H., 1854, *Guidebook Through the Royal Dockyard*, Portsmouth: Henry Lewis.

Lipscomb, F., 1967, *Heritage of Sea Power*, London: Hutchinson.

Lunn, K., and Day, A. (eds), 1998, *Inside the Wall: Recollections of Portsmouth Dockyard 1900–1950*, Portsmouth: University of Portsmouth.

Lunn, K., and Day, A. (eds), 1999, *History of Work Labour Relations in the Royal Dockyards*, London: Mansell.

MacDougall, P., 1982, *Royal Dockyards*, Vermont: David and Charles.

MacDougall, P., 2012, *Chatham Dockyard: The Rise and Fall of a Military Industrial Complex*, Stroud: History Press.

McCart, N., 1999, *Tiger, Lion and Blake 1942–1986*, Cheltenham: Fan.

McGowan, A., 1999, *HMS Victory, Her Construction, Career and Restoration*, London: Chatham.

Morriss, R., 1983, *The Royal Dockyards during the Revolutionary and Napoleonic Wars*, Leicester: Leicester University Press.

Oppenheim, M., (ed.), 1896, *Inventories of the Reign of Henry VI, 1485–1488 and 1495–1497*, London: Navy Records Society.

Patterson, B., (ed.), 1984, *The Royal Dockyard at Portsmouth in 1929*, Portsmouth: Portsmouth Dockyard Historical Society.

Patterson, B., 1989, *Giv'er a Cheer Boys: The Great Docks of Portsmouth Dockyard*, Portsmouth: Portsmouth Royal Dockyard Historical Society.

Patterson, B., 1991, *Windows on the Past*, Portsmouth: Portsmouth Royal Dockyard Historical Society.

Patterson, B., 1998, *Ships In and Out of Portsmouth Dockyard*, Portsmouth: Portsmouth Royal Dockyard Historical Trust.

Patterson, B., 2008, *Portsmouth Dockyard Chronology*, Portsmouth: Portsmouth Royal Dockyard Historical Trust (see also Dockyard Timeline, http://portsmouthdockyard.org.uk/chronology/history-of-the-dockyard, accessed 8.3.17).

Preston, A., and Major, J., 2007, *Send a Gunboat*, London: Conway.

Riley, R. C., 1976, 'The Industries of Nineteenth Century Portsmouth', *Portsmouth Papers*, No. 25, Portsmouth: City of Portsmouth.

Riley, R., 1985, 'The Evolution of the Docks and Industrial Buildings in Portsmouth Royal Dockyard 1698–1914', *Portsmouth Papers*, No. 44, Portsmouth: City of Portsmouth.

Roberts, J., 2013, *The Battleship Dreadnought*, London: Conway.

Rodger, N.A.M., 1997, *Safeguard of the Sea*, London: Harper Collins.

Rodger, N.A.M., 2004, *Command of the Sea*, London: Allen Lane.

Sadden, J., 2001, *Portsmouth, Heritage of the Realm*, Chichester: Phillimore.

Sadden, J., 2012, *Portsmouth and Gosport at War*, Stroud: Amberley.

Slight, H. and J., 1828, *Chronicles of Portsmouth*, London: Lupton Relfe.

Sturton, I. (ed.), 2008, *Conway's Battleships*, London: Conway.

Tracy, N. (ed.), 2002, *Age of Sail*, London: Conway.

Thomas, R., and Patterson, B., 1998, *Dreadnoughts in Camera*, Stroud: Sutton Publishing.

Warner, J., 2004, *John the Painter*, London: Profile.

Webb, J., 1971, *An Early Nineteenth Century Dockyard Worker*, Portsmouth: Portsmouth Museums Society.

Wells, J., 1987, *The Immortal Warrior*, Emsworth: Kenneth Mason.

Winfield, R., 2005, *British Warships in the Age of Sail, 1793–1817*, London: Chatham.

Winfield, R., 2009, *British Warships in the Age of Sail, 1603–1714*, London: Chatham.

Winfield, R., 2014, *British Warships in the Age of Sail, 1714–1792*, Barnsley: Seaforth.

Winfield, R., 2014, *British Warships in the Age of Sail, 1817–1863*, London: Chatham.

Winton, J., 1989, *The Naval Heritage of Portsmouth*, Southampton: Ensign.

Wragg, D., 2006, *The Royal Navy Handbook 1914–1918*, Stroud: Sutton.

INDEX